Mrs. Basley's Poultry Book
1001 Questions on Up To Date Poultry Culture

by Mrs. A. Basley

with an introduction by Jackson Chambers

THE WORLD'S LARGEST SELECTION OF VINTAGE POULTRY BOOKS

www.VintagePoultry.com

Self Reliance Books

Get more historic titles on animal and stock breeding, gardening and old fashioned skills by visiting us at:

http://selfreliancebooks.blogspot.com/

Introduction

I am pleased to present yet another title on Poultry.

The work is in the Public Domain and is re-printed here in accordance with Federal Laws.

As with all reprinted books of this age that are intended to perfectly reproduce the original edition, considerable pains and effort had to be undertaken to correct fading and sometimes outright damage to existing proofs of this title. At times, this task is quite monumental, requiring an almost total "rebuilding" of some pages from digital proofs of multiple copies. Despite this, imperfections still sometimes exist in the final proof and may detract from the visual appearance of the text.

I hope you enjoy reading this book as much as I enjoyed making it available to readers again.

Jackson Chambers

INTRODUCTION

In the hope of helping beginners and others of my friends in the poultry business, and in response to urgent requests for a book on poultry culture from my pen, I offer this little volume.

It is a synopsis of many chapters of my "Woman's Work in the Poultry Yard" and other talks on poultry, and embodies the personal, practical experiences I have been through myself in many years of pleasant work in the poultry yard. Its object is not necessarily to urge anyone into the business, but to encourage and help beginners and especially newcomers on the Pacific Coast, where conditions differ materially from those in the East and where there is an increasingly large demand for both poultry and eggs; where the poultry business is about as profitable as any that can be undertaken and a good living may be made in the pure air and sunshine by any industrious man or woman.

Having for many years been lecturer at the Farmers' Institutes in the Extension Courses of the University of California and having been editor or associate editor of four agricultural and other newspapers on the Pacific Coast, many questions have during this time been propounded to me relating to the poultry business, its difficulties, the troubles of poultry raisers and the ailments of fowls. Some of these thousand and one questions will be found in this book with the answers to them, also remedies for the diseases or ills of fowls in this climate.

Hoping and feeling sure that my little book may prove a help to all its readers, I am

Very cordially your friend,

Mrs. A. Basley

MRS. A. BASLEY

TABLE of CONTENTS

CLASSIFIED INDEX

Common Sense Poultry Houses

The poultry business is one of the most fascinating as well as the most profitable, considering the amount of capital invested, in the West. The conditions here, however, differ so greatly to those in the East and other localities that the ways of treating the fowls must also be different. The needs of fowls do not vary, the resources of the places do and the success of the poultry raiser greatly depends upon adapting the conditions of the locality to the need of the fowls.

Nothing is more important than the proper housing of chickens. The style of house a man builds for his birds will depend upon his means and inclinations. It is not always the most expensive house that gives the most eggs. In planning poultry houses and yards two or three principles should be firmly held in mind: First, the house must have a liberal supply of oxygen, which can only be supplied by perfect ventilation; secondly, it must be free from draughts and be dry, and, thirdly, be easily accessible to the attendant not only for cleaning and spraying but to enable one to handle the fowls when on the perches. It should also be large enough to avoid crowding of the fowls.

The laying hens should be kept in yards in permanent houses, easy of access, whilst the young and growing fowls will do best on free range with movable houses, called sometimes colony houses. These give the best results.

After many years of experience here the writer has found that there are two classes of houses admirably adapted to the needs of the fowls and to this climate. These are called the open front or the "fresh air" house and the "mushroom" house. What is meant by an open front house, is a house enclosed on three sides and roof with one side open to the fresh air. This style house can be con-

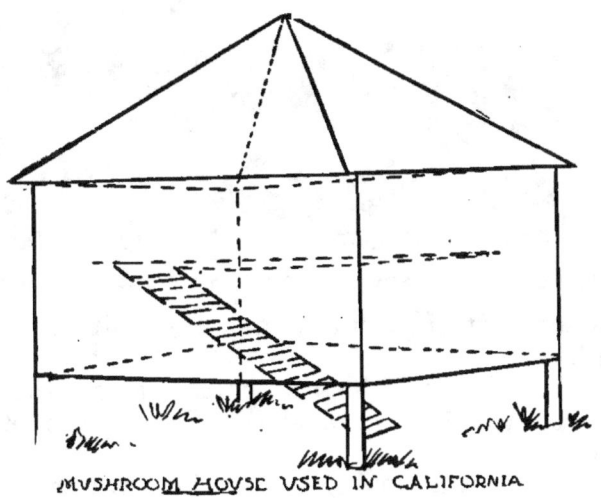

MUSHROOM HOUSE USED IN CALIFORNIA

structed as a separate and movable house or as a continuous and scratching shed house. A plain open front house without a scratching shed attached, is used in many places as a colony house where fowls have free range or where they are kept in an orchard.

The "mushroom" house is built tight on four sides and roof, without any floor and is raised from the ground about twelve inches.

Cuts of both of these styles of houses will serve to show their construction.

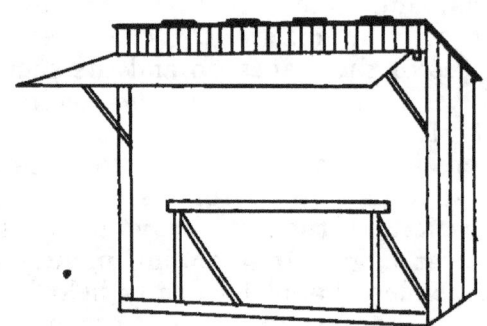

Open Front House Without Scratching Shed.

A "fresh air" house that proved excellent and which I used for years on my ranch was one hundred and twenty feet long and ten feet wide. It was divided into six houses with scratching pens. I also had another which suited me well. It was eight feet wide and a hundred fet long; besides that I had twenty colony houses for the young and growing stock, and two brooder houses.

The continuous house and scratching shed of which I give a

View of Mrs. Basley's Continuous Fresh Air House and Scratching Shed.

photograph and part of ground plan were built of flooring, tongued and grooved.

The other house was of boards, battened, and the colony houses of resawed red wood or of shakes. Some were of rubberoid or building paper.

Many of the artistic looking house plans which may be found in poultry books were planned by men who never owned a chicken, and if built in this, or in any other climate, would be highly unsatisfactory. The plans here described have all been used either by myself or by successful poultry raisers. I have seen them all and can assuredly recommend them for use on the Pacific Coast.

The houses I am describing are of the inexpensive kind, for so great is the variety of plans of houses designed for fowls that it would be impossible to mention them all in a short article. We will, therefore, consider only a few of the cheapest and most satisfactory small houses adapted to this climate.

The first requisite in the house is pure air. To secure this the

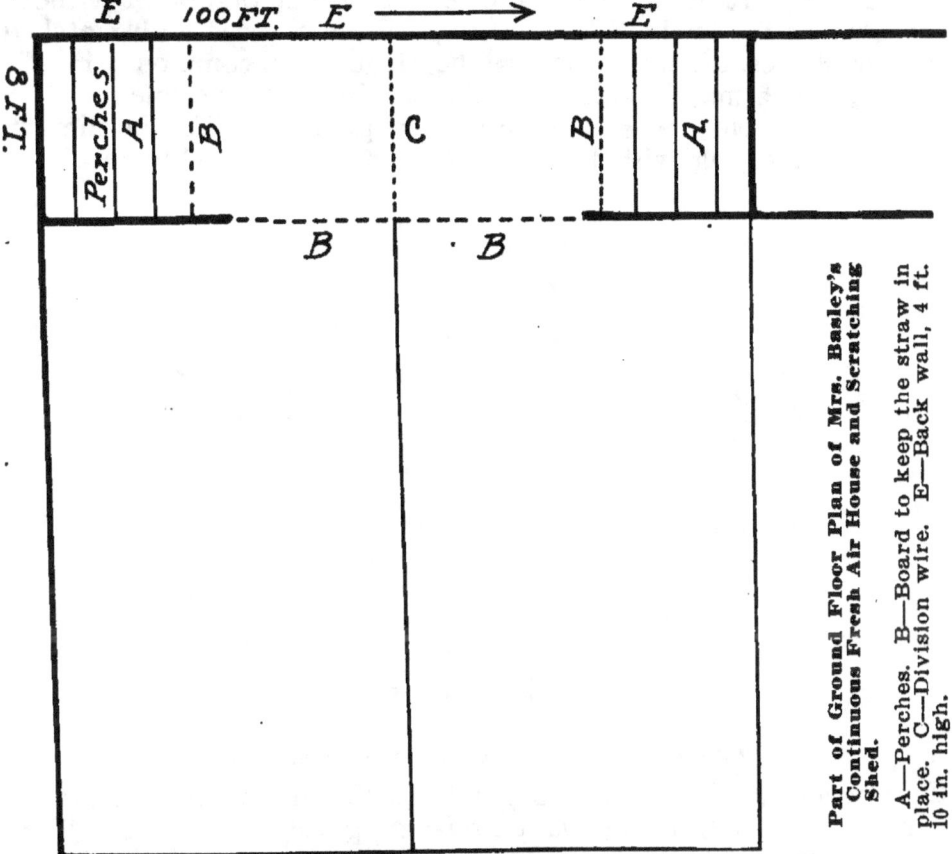

Part of Ground Floor Plan of Mrs. Basley's Continuous Fresh Air House and Scratching Shed.

A—Perches. B—Board to keep the straw in place. C—Division wire. E—Back wall, 4 ft. 10 in. high.

ventilation must be at the bottom. Some people think that the bad air ascends, but this has been proved a mistake,—the foul gases descend; the pure air and the warm air are lighter and they rise and we want to keep them in, but if we have an opening for ventilation at the top or near the top of the house we lose the warmth. A loss of warmth at night in the winter means a loss of eggs, or more food is needed to supply this loss. The ventilation should either be at the bottom, or one entire side of the house should be left open.

A Variety of Houses

The accompanying rough little cut of a "mushroom" house will give some idea of the bottom ventilation. Houses like this were

used by a successful poultry man. He made a light frame five feet
square and five feet high. This he covered with canvas and the
roof he made of rubberoid roofing. He left a space below of ten or
twelve inches. These "mushroom" houses were tipped over every
day to be sunned or cleaned. I improved upon his plan by making
a door of one whole side, for I wanted to be able to handle my fowls
at night without tipping the house over. Perches should be placed
about twelve inches above the open space, and in the case of heavy
breeds a small ladder or run board should be placed for them to
reach the perches easily when going to roost. The advantages of
such a house are its lightness, and the free circulation of air without
draughts on the fowls. These houses can be covered with matched
lumber, shakes, canvas, burlap, rubberoid, or even common domestic
muslin, which may be oiled or painted with crude petroleum.

The open front house is admirably adapted to California climate.
It is now meeting with favor even in the rigorous climate of the

Holbrook's Canvas Covered Mushroom House.

East, where poultry raisers begin to realize the value of fresh air
without draughts, if they want to have vigorous hens that will lay
eggs in the winter time. I have been using the open front houses
of various sizes for over twelve years and can assert that they are
the only kind I ever want to use. Another style open front house
that I have seen and like very much is fifteen feet by eleven feet six
inches, and is seven feet high at the back and four feet at the open
front. It is constructed of rubberoid or malthoid and is almost
vermin proof. It is divided in the middle by chicken wire, so forming either one house or two as required. The roof is first covered
with two-inch chicken wire to support the rubberoid. At the bottom
of the walls next to the ground it is boarded up for about two feet
all the way round; this is to keep in the straw, for all the floor
space of the house is used as a scratching pen. The sides and back
above these boards are made of panels of rubberoid nailed to light
frames without the chicken wire. These panels are taken down on

all fine days to sun and air the house. The panels are kept in place by large wooden buttons. The front is entirely open or only closed by chicken wire except when it rains, then a burlap curtain is let down. The perches are near the back of the house about six inches above the dropping boards. The dropping boards are made of the rubberoid on frames. They are four feet wide and are placed on cleats two feet from the floor. This is a double house and each side will hold from twelve to twenty hens. The above description is of the Hoffman house pictured below.

A cheap and substantial house can be made of two piano boxes. The simplest way to make such a house is as follows: Removing the backs of the piano cases, place the cases back to back thirty inches apart, on light sills. Use the boards which were the backs to fill up the thirty inches on the sides and roof; cover the roof with rubberoid or with oil cloth and you have a comfortable house, that will hold about a dozen or twenty hens, at a small cost. The front of the piano box house should either be hinged so it can

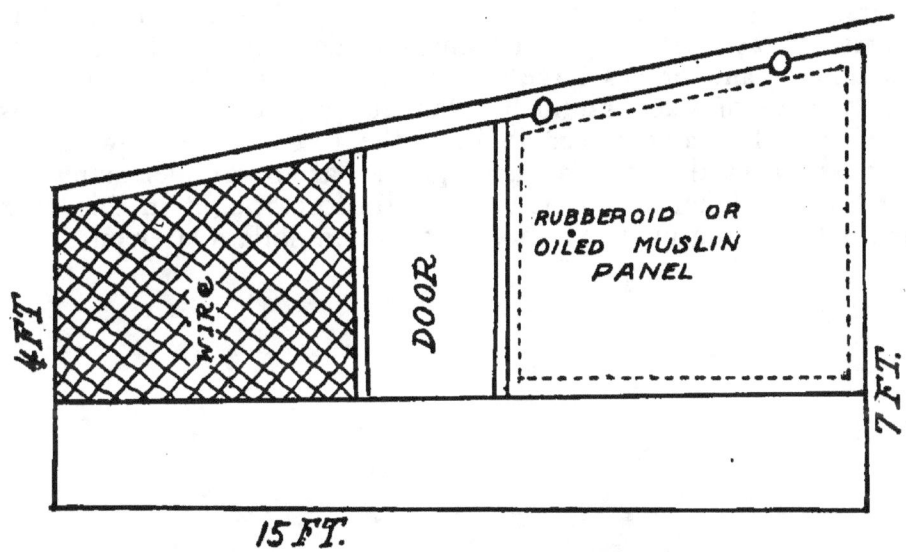

Hoffman's Combination Open Front House and Scratching Pen.

always be kept open except during the rain or it may be entirely dispensed with and a burlap curtain used to keep out the rain. The cost of this piano box house is about three dollars.

Inexpensive Colony Houses

An inexpensive colony house is pictured on page 10. This house is of resawed redwood, four by six feet. It is light and easily moved. The front is on hinges and it is always kept open except during rain, and when it is closed it only comes down six inches below the perches, leaving an open space of about fifteen inches across the entire front.

Still another style of colony house and one well adapted for use in an orchard or in the colony plan has been in use for some years on a large poultry ranch in California. The house is eight by ten feet and two feet to the eaves; all the framework, including the runners, is of two by three inch stuff, and the walls and ends are of one by twelve inch boards, shiplapped so as to avoid using battens. The rafters are five feet four inches long, and three pairs are used; a one by six inch strip is run all around the outside of the roof to form the eaves and also to make it tight; eight pieces of one by four are used for sheathing, and the sawed shakes are close so that there is no draught from that source; the only opening is from the front which is open at all times. The houses do not require cleaning, for they are on runners, and are slid along about fifteen feet each time. Thus they are on fresh ground and much cleaner than one could do it in any other manner.

Painting the Houses

For painting the houses I have found nothing better than the crude petroleum. I add to it for all my houses, red Venetian paint mixed with a little kerosene or distillate oil, to thin it. This colors them a handsome chocolate. Creosote stain of a dark green in also a very good color, harmonizing well with the landscape, and both of these are preventive of mites and keep their color well for several years. A good whitewash also is quite suitable. The color is a matter of taste after all, and I am only describing the inexpensive methods I and others have successfully used. The whole plant, irrespective of size, should be planned symmetrically; the houses made all alike and placed in line; the large in one row and the smaller in another and all arranged so as to save as many steps for the care-taker as possible. A little forethought in this matter at the beginning may save many steps and dollars later on.

Week's Portable Canvas Houses.

What Variety to Choose

"Poultry for profit" is the slogan. We are all looking more or less for the "almighty dollar." Every week, almost every day, I am appealed to for information as to which breed is the most profitable. I can and often do tell which breed I have found the most profitable in the twenty years I have bred, but I cannot decide for another person what his or her likes or dislikes may be, nor can I tell what poultry will suit another's location or market. That, each one must decide for himself or herself, and then get the best of that breed to start with.

A hint as to what to start with may help some of our readers. First of all study your market, decide whether it requires a brown

Barred Plymouth Rock Cockerel

or a white egg and choose accordingly; secondly, decide what you will do with the surplus chickens, although this may seem like counting the chickens before they are hatched. Will you sell them as broilers and fryers or use them as roasters or capons? Thirdly, it is always a good plan to look ahead and choose a breed with a prospective value and demand—one of the breeds that may be rare in your neighborhood, or one of the newer breeds, such as

Silver Laced Wyandotte Hen.

the Orpingtons, Columbian Wyandottes or Favorelles. Choose a breed for which there is likely to be a large demand for eggs for hatching and for breeding stock. Or else take one of the best old breeds that you know will make you money from the start. Whatever breed you decide upon get the best of that breed, and from a reliable breeder.

Different Breeds

A brief review of the different classes and breeds of domestic fowls may be of use to beginners. There are a large number of breeds in this country suitable to any branch of the business, with all colors of plumage and size. Some especially adapted to the farm, others to closer confinement, as on the city lots, and still others—like the beautiful little bantams—adapted to lawns and front yards.

The American Class

The American class consists of what are called the dual-purpose fowl. That is, they are good for market as well as excellent layers, so when their day of usefulness in the egg basket is over they can end their existence on the table. This class gives us the Barred, Buff and White Plymouth Rock, the Silver, Golden, White, Buff,

Mrs. Basley's White Plymouth Rock 'Snow Queen.' Layed 225 Eggs in 9 Months

Silver Pencilled, Black, and Columbian Wyandottes, the Single and Rose Comb Rhode Island Reds, the Buckeyes, the Black, White and Mottled Javas, and the American Dominique. Of the list no doubt the Barred Plymouth Rock is the best known and most popular; it may be said to lead the American class. Next to it in popularity is the White Plymouth Rock. This breed led in numbers at a late show in Madison Square Garden in New York, which is a sure indication of its popularity. The order of the rest might be given as follows: White Wyandotte, Rhode Island Reds, Buff Wyandotte, Buff Plymouth Rock, Silver Wyandotte, Partridge Wyandotte, Golden Wyandotte, Buckeyes, American Dominique, Black Java.

The standard weights of the above are as follows: All of the Plymouth Rocks, cock, 9½ pounds; cockerel, 8 pounds; hens, 7½ pounds, and pullets, 6½ pounds. All of the Wyandottes, cock, 8½ pounds; cockerels, 7½ pounds; hens, 6½ pounds, and pullets, 5½ pounds. The Rhode Island Reds, cock, 8½ pounds; cockerel, 7½ pounds; hen, 6½ pounds; pullet, 5 pounds. Buckeyes half a pound heavier except pullets. The Javas are of the same weight as the Plymouth Rocks, and the American Dominiques, cock, 8 pounds; cockerel, 7 pounds; hen, 6 pounds; pullet, 5 pounds.

The Mediterranean Class

In the Mediterranean class, we have the Single and Rose Comb Brown, Single and Rose Comb White, Black, Buff and Silver Duckwing Leghorns; the Black and White Minorcas; the Blue Andalusians, the Black Spanish, and Mottled Anconas.

The Mediterranean class is particularly well adapted to the cli-

A Flock of A. H. Memmler's Columbian Wyandottes.

mate of California, which greatly resembles that of their home in the old countries.

In point of popularity and merit, the kinds might be classed as follows: White Leghorn, Brown Leghorn, Black Minorca, Blue Andalusian, Black Spanish, Rose Comb Brown Leghorn, Rose Comb White Leghorn, Buff Leghorn, White Minorca, Anconas, Silver Duckwing Leghorn and Black Leghorn. The Black Minorca, White Leghorn and Black Spanish give the largest sized eggs.

All of the Mediterraneans have white shelled eggs. There is no standard weight to the Leghorns. They are small birds, weighing 3 or 4 pounds. Of the Black and White Minorcas, the cock weighs, 9 pounds; cockerel, 7½ pounds; hen, 7½ pounds; pullets, 6½ pounds. The weights of the Andalusians are, cock, 6 pounds; cockerel, 5 pounds; hen, 5 pounds; pullets, 4 pounds.

The Black Spanish weights are, cock, 8 pounds; cockered, 6½ pounds; hens, 6½ pounds; pullets, 5½ pounds. These lay an extra large handsome white-shelled egg.

The Blue Andalusian has the unique distinction of wearing the national colors—red, white and blue—its plumage being blue, its face and eyes red and its ear-lobes white.

The Asiatic Class

The Asiatic class consists of the Light and Dark Brahmas, White and Black Langshans, the Buff, Partridge, White and Black Cochins. In point of popularity, they would be about in this order: Light Brahmas, Black Langshans, Buff Cochins, Partridge Cochins, Dark Brahmas, White Cochins, White Langshans and Black Cochins. The standard weights are: Light Brahmas, cock 12 pounds, cockerel 10 pounds, hen 9½ pounds, pullet 8 pounds.

Weights for Dark Brahmas are: Cock 11 pounds, cockerel 9

White Plymouth Rocks.

pounds, hen 8½ pounds, pullet 7 pounds; Buff, Partridge and White Cochins, cock 11 pounds, cockerel 9 pounds, hen 8½ pounds and pullet 7 pounds; Black and White Langshans, cock 10 pounds, cockerel 8 pounds, hens 7 pounds and pullet 6 pounds. The eggs of all of the Asiatic class are a dark brown.

The English Class

The English class is composed of the White, Silver-gray and Colored Dorkings, the Red Caps and the Buff, Black, White, Spangled and Jubilee Orpingtons in both single and rose combs. The White Dorking weighs as follows: Cock 7½ pounds, cockerel 6½ pounds, hen 6 pounds, and pullet 5 pounds; Silver-gray Dorkings, cock 8 pounds, cockerel 7 pounds, hen 6½ pounds and pullet 5½ pounds; Colored Dorkings, cock 9 pounds, cockerel 8 pounds, hen 7 pounds and pullet 6 pounds; Red Caps, cock 7½ pounds, cockerel 6

Goodacre's Black Orpington Cockerel "Royal Arms."

pounds, hen 6 pounds and pullet 5 pounds; Orpingtons, cock 10 pounds, cockerel 8½ pounds, hen 8 pounds and pullet 7 pounds.

The French Class

The French class is composed of the Houdans, Crevecoeurs, La-Fleche and Favorelles. The Houdans weigh, cock 7 pounds, cockerel 6 pounds, hen 6 pounds and pullet 5 pounds; the Crevecoeurs, cock 8 pounds, cockerel 7 pounds, hen 7 pounds and pullet 6 pounds. The Crevecoeurs and La Fleche are favorites in France, but are rarely found in this country, as they are not popular in the market here on account of their dark colored shanks.

The Hamburg Class

The Hamburg class is composed of most excellent layers, of white eggs. They are the Silvered Spangled, Golden Spangled, Silver Penciled, Golden Penciled, White and Black Hamburgs and the Silver and Golden Campines. No weights are given for the Hamburgs and Campines.

The Polish Class

The Polish are more of a fancy fowl. They are the White Crested Black, Golden, Silver, White, Bearded Golden, Bearded Silver,

Houdan Cock Heading C. W. Bessey's Pens.

Bearded White and Buff Laced.. They lay white eggs; no weights are given in the Standard for them

The Game Class

In the Game class, we have the Black Breasted Red, Brown Red, Golden Duckwing, Silver Duckwing, Red Pyle, White, Black and Birchen Games, Cornish and White Indian Games, Black Sumatras and Black Breasted Red Malays.

The standard gives no weight for Games, excepting for Indian Game, viz., cock 9 pounds, cockerel 7½ pounds, hen 6½ pounds and pullet 5½ pounds; Malays, cock 9 pounds, cockerel 7 pounds, hen 7 pounds and pullet 5 pounds.

Turkeys

The most popular variety of turkey is the Bronze, then comes the White Holland, another splendid variety. Among others we have the Black, Buff, Bourbon Red, Slate, Narragansett and Wild.

The weights for Bronze are cock 36 pounds, yearling cock 33 pounds, cockerel 25 pounds, hen 20 pounds and pullet 16 pounds; for White Holland, cock 26 pounds, cockerel 18 pounds, hen 16 pounds, pullet 12 pounds.

Ducks

The Pekin is "The American Duck" with its white plumage and heavily meated body. Their weight is as follows: Adult drake 8 pounds, young drake 7 pounds, adult duck 7 pounds, young duck 6 pounds. Another white variety, very popular in England, is the

A Pair of Beautiful Bronze Turkeys.

Aylesbury. Weight for adult drake 9 pounds, young drake 8 pounds, adult duck 8 pounds, young duck 7 pounds. The colored Rouen have similar weights and plumage to the Wild Mallard, the drakes having bright green heads. Other popular varieties are the Indian Runners, both colored and white, called the Leghorn of the duck family, being rather small, very active and immense layers of fine white eggs Then there are the Buff Orpington Ducks— becoming very popular; the Blue Swedish, Black Cayuga, Colored and White Muscovy, Call and Black East India, these latter being more ornamental varieties.

Geese

Perhaps the easiest kept and noisiest of all our large variety of domestic fowl are geese, and where conditions are suitable they prove very profitable. The Toulouse, a large gray variety, and the White Embden, seem the most popular of the pure bred varieties, and the weights for either variety are, for adult gander 20 pounds, young gander 18 pounds, adult goose 18 pounds, young Toulouse goose 15 pounds and Embden young goose 16 pounds. Other varieties are the African, Brown and White Chinese, Canadian and Egyptian; these are either used for ornamental purposes or for crossing.

Selection of Breed

Knowing the values and weights of the different standard breeds the beginner will be enabled to make his choice, and have no trouble in finding the proper selection.

Supposing egg production is the principal object, the beginner will have to decide according to the demand of his nearest market. Boston requires brown eggs, San Francisco white eggs, while Los Angeles seems to be content with either. If you are living near

Indian Runner Drake   Buff Orpington Ducks

San Francisco one of the Mediterranean breeds will prove the most valuable to you. The Minorcas, Black Spanish and some of the strains of White Leghorns lay the largest and finest looking eggs. One correspondent who asks for justice for the Minorcas says he has Minorca hens which lay eggs weighing nearly three ounces, and there were Leghorn eggs on exhibition in a late poultry show which weighed five eggs to the pound, but these were from hens "bred to lay." The Brown Leghorns and Hamburgs give many eggs—white eggs also—but smaller, which is an objection in a good market. Should broilers be the object, we should choose the White Wyandottes or White Plymouth Rocks. These latter are exceptionally fine winter layers. For roasters and capons, the Light Brahmas or any of the Plymouth Rocks are the favorites. If two breeds are wanted, we should personally prefer the White Leghorns and White Plymouth Rocks. The White Plymouth

Rocks will give the winter eggs and the White Leghorns the spring and summer eggs in great abundance, although they may not lay as many eggs in the winter as the White Rocks. In the early spring the White Rock eggs can be set for early broilers and roasters, while the Leghorns are doing their heaviest laying, and in April and May the Leghorn eggs can be set for the following season's eggs. In this manner there will be a constant succession of eggs for market, and broilers and roasters in season. Always having something to sell means a regular income. Something to market at least once a week. A poultry and egg route and the reputation of having none but the choicest goods to offer is the secret of success.

Silver Campines.

Eggs for Breeding

Having chosen the breed which suits us best, let us talk on how to get the most out of that breed, for I think we are all agreed that if we keep poultry for profit, we want to make as much as we can out of it. Therefore, having got our fowls, we must treat them right. The natural instinct of a fowl is to make a nest for itself and raise a family of its own in the spring time. It never considers its owner's profit or loss, therefore to make it answer our purpose, to develop it into a money-maker for us, we must either change its nature or deceive it. We must let it imagine that it is the time of year for nest making and family raising. We must supply it with the conditions of springtime. Our own lives are artificial and the conditions surrounding our domestic hens are also artificial, but we must, if we want success, copy as far as possible Nature's ways with fowls and follow Nature's plans.

In the spring not only do we want egg production, but we want good, strong fertility in our eggs. We want fertile eggs now, for are we not prearranging to have plenty of vigorous pullets to lay those high-priced market eggs next fall? Are we not anticipating sturdy cockerels to win prizes at next winter's shows, or to make toothsome frys or delicious roasts?

Fertile eggs are now in order. How shall we get them? First we must have vigorous and healthy parent birds; we usually have healthy birds in the spring of the year, for the moult is well over and the ailments which prevail in the fall—colds, catarrh and sore throats, all classed as roup—have yielded to treatment, or the victims are no more. The chickenpox, which also is a fall disease, has about disappeared, and the birds are in good condition.

Eggs for Breeding, Packed Correctly for Shipment.

Vigor is Necessary

Vigor is the first requisite for fertile eggs. To have vigor, the hens must have exercise; every grain they eat should be scratched or dug out of the straw or litter in their scratching pen. A hen that is very fat—over-fat—will not have fertile eggs and will not have strong, sturdy chickens. It is neither kind nor wise to over-fatten your breeding hens, but they must be fed the proper food for fertility. How can we decide what food to feed for fertility? Let us interrogate Nature again. The wild bird, the Gallus Bankiva from which sprung all our domestic fowls, lays her eggs and raises her young only in the spring. She only has two broods of about thirteen eggs each, but those eggs are rarely infertile. What does she eat? Principally insects and the tender green grasses or small leaves, not much grain, for the seeds have fallen and have begun to sprout and grow.

During the winter Nature has supplied the birds with grains in plenty, so they have put on fat to withstand the cold; but now there are only a few grains left and the fowls are becoming thinner, yet Nature does not starve them, only gradually changes the ration and gives them worms and larvae, insects of all kinds, for the insect life has also commenced to pulsate and develop; the buds are bursting, too, and the tender green appears and beautiful spring is here, providing all the green food they can eat. How about our captive hens? In our bare back yards, with only the ration we choose to give them? Poor things; they have a natural craving for the tender green, a wild desire for the succulent insect or animal food! See, how they will fight over or scramble for the meat that is thrown to them, or for the head of lettuce! They try to tell us in their own way what they require to produce fertile eggs at this season of the year.

How to Feed

How shall we follow their teachings? Increase the amount of their animal food and give the breeding fowls more green food. How shall we do this? Increase gradually whatever animal food we are now feeding until from 20 to 30 per cent of their daily food is animal food. The best animal food is fresh meat of some kind; the scraps and bones left over at the market; this ground or chopped finely is the best I know of. Rabbits, squirrels, gophers, are all good fresh meat. If fresh meat cannot be obtained, you can get at the poultry supply houses granulated milk, dried blood, blood and bone, beef-scrap and other animal food. The best green food is fresh-cut clover lawn clippings, green alfalfa, lettuce, cabbage and other vegetables.

The Male Bird

The male bird is considered as half the pen. The germ or seed of life of the future chicken is from the male. Be sure to have the male vigorous and healthy, and see to it that he gets sufficient food of the right quality. The male bird is often so gallant that he calls up his wives and they greedily eat all the best part of the food, choosing first the meat or animal part, which is the most necessary for fertility, and the husband and father of future chicks, on which so much depends, is half starved, becomes thin and light. Every male bird when being used to fertilize eggs should be fed extra, either in a pen or corner by himself, or out of your hand at least once a day.

Mating

In mating up the pens I have found the most satisfactory number to mate is about eight or not over ten females of the American breeds to one male. From twelve to fifteen of the Leghorns or Mediterranean birds, and from six to eight of the Asiatic class to one male. Some breeders advocate using two male birds in one pen, alternating them day about, or three male birds for two pens, allowing one bird to rest every second or third day. I never did this, because I was pedigreeing my fowls, and I never found any necessity for it.

Caring for Fertile Eggs

Having the fertility assured, the next thing is to take care of the eggs from the time they are laid until incubation begins. Eggs should be kept in a moderately cool, quiet place; not in a draught. I always imitated Nature and turned the eggs, just as a hen would, every day, keeping them in a box either in the cellar or a large, dark, but airy, closet. Some people keep them in fillers with the little end down, but I prefer following Nature's ways and leaving them on their side.

To Choose Eggs for Hatching

To choose the eggs for hatching I use an egg tester or I roll up a copy of the Live Stock Tribune in the shape of a telescope, putting the egg at one end in the sun and my eye at the other end. If the egg shell is speckled or thin at one end, or has thin blotches on it or is misshapen in any way or if it feels chalky to the touch I reject that egg, relegating it to the kitchen, for these eggs will not hatch. I also reject very small eggs, as they are laid by pullets or by overfat hens and if they hatch the chickens will be weaklings. The very large eggs should also be rejected as they may have double yolks and these seldom hatch healthy chickens. Above all, never sell for hatching eggs those as described above. The best eggs are the egg-shaped eggs, with good, firm, smooth shells and not narrow waisted.

Buff Cochin Hen.

Views of Ross & Tate's Orpington Reservation.

Eggs for Market

The hen in her wild state lays about thirty eggs per year. The farmer's average hen lays not over one hundred. On egg farms the average is 150 and some of the fowls of the "bred to lay" strains will average even more.

There are 365 days in the year and I do not see why a pullet that is fully matured, that comes from an egg-laying strain, a pullet properly fed and cared for should not lay over 200 eggs per year; in fact, I have had hens that will do even better than that. I will admit that a hen will not lay 200 eggs a year without constant and intelligent care, and the question 'confronting us is, will the additional number of eggs pay for this care? Also how shall we give this care and secure these results?

You hear of heredity and pedigree in cows, in horses, in dogs. Heredity is as important with hens as with any other stock. Heredity has as much to do with the success of hens as the right handling.. Heredity (or pedigree) and handling must go together. The two-hundred-egg hen must be "bred to lay." She must come from an egg producing family. No matter how scientifically a hen is fed, or how well housed, you cannot make an extra fine layer out of one whose parents for generations past have been poor layers. It is impossible to take a flock of mongrels and scrubs and get 200 eggs each a year from them, although good handling will greatly increase the yield of even mongrels.

The different breeds require different handling but no matter what breed you have, there are three essentials to egg production—comfort, exercise and proper food.

Comfort

Under the head of comfort comes first of all cleanliness. A hen that has lice, or fleas, or mites, or ticks on her can not lay her full amount of eggs. You must help the hen in her efforts to make you money. Give her every encouragement to lay. Cleanliness everywhere. A comfortable, enticing nest, rather dark, where she may

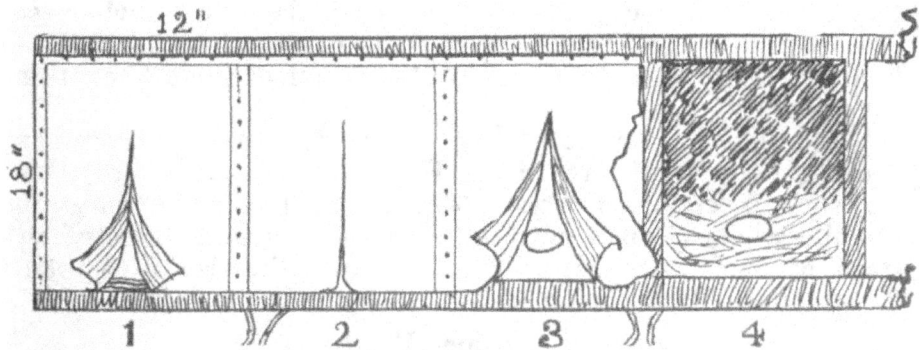

Nest to Keep Hens from Eating Eggs.
1.—Canvas curtain slightly turned back.   2.—Curtain concealing nest.
3.—Curtain thrown back to show construction. 4.—Curtain removed.

stealthily deposit her precious egg. Renew the nice clean straw once a month. Do everything to coax the hens to lay. If trap-nests are used, there should be enough of them so that the hens will not be kept waiting, for by keeping a hen off the nest she will frequently retain her egg until the next day, and will soon learn to be a poor layer. Cleanliness means a clean, sweet-smelling roost-ing place, where she may sleep undisturbed by lice or mites. Just think for a moment how in the human family a fresh, clean bed in a quiet room will court slumber. I have passed the night in an Arab's tent in Africa that was infested with fleas, and my heart is full of sympathy for a hen that has to live in some of the mite-infested henneries I have seen in California. Under this head comes freedom from draughts. A draught in this country will give hu-man beings face ache, neuralgia, earache and a swelled face. It has exactly the same effect on hens. Influenza, swelled head, roup, al-ways or almost always commence from a draught (combined with lice). Comfort means also pure, fresh air without any draught, and pure, fresh water to drink.

Exercise

You know how in the human family exercise is recommended. Physical culture, gymnastics, Ralston exercises, Swedish move-ments, fencing, etc., and those who may be too feeble to exercise for themselves, pay others to rub, pound and knead or massage them to get the same effect.

Exercise is as necessary for the hen as for the human being and more so, for the hen's exercise of scratching develops the egg pro-ducing organs and strengthens them, and hens which exercise lay many more eggs than lazy hens. If you have a vigorous scratcher among your hens you may be sure she is a good layer.

Exercise a hen must have to develop the egg-making organs. She absolutely must scratch if she is to make a living for herself and you. I consider a scratcing pen as necessary for hens in con-finement as food. My scratching pens were twelve or fifteen feet long and eight feet wide, but in small yards I have made very satis-factory little pens by nailing four boards six feet long together, forming a square. The boards should be twelve inches wide and the pen filled with wheat straw or alfalfa hay or any good litter. I do not like barley straw on account of the beards, which some-times run into the hen's eyes, nostrils or mouth and cause death. Foxtails, burr clover and wild oats are all dangerous on this ac-count.

I feed all the grain scattered over the straw and my hens scratch and dig happily all day long. The straw or hay is soon broken into short pieces and fresh straw must be added about once a week and the whole cleaned out and used for mulching trees when the straw becomes dirty. It will depend upon the size of the pen and the number of hens using it. .

Proper Food

What it is and how much to give. The scientists tell us that the proper food or the "balanced ration" is composed of one part of protein to four parts of carbo-hydrates. Before discussing this "balanced ration" let us interrogate Nature and find out how a hen balances her own ration.

Let us take a hen as she comes in from foraging in the fields after a long day in summer. Let us kill her and examine her crop. What do we find? Grains of wheat, barley, corn, according to where her rambles have led her; bits of grass, clover and vegetables; some bugs, worms and grasshoppers; here and there a bit of gravel and a lot of matter partially digested that we can not recognize. The first thing that impresses us is that the hen likes variety, and the second thing that this variety consists of animal food (bugs, worms, insects) grains and green food. This is the "balanced ration" balanced by the hen herself to suit her needs in the summer time when eggs are plentiful. If we want eggs in the winter we must, as far as possible, give the same conditions, the same variety of foods, with plenty of pure, fresh water, never forgetting that about seventy per cent of the egg is water.

But to return to the "balanced ration." We know that a hen

Beautiful White Eggs Layed by A. H. Wheeler's Black Minorcas.

requires a certain amount of food to keep her alive and thriving; above that the surplus goes either to making the egg inside her or to making fat.

The hen is an egg-making machine but if you put into that machine none of the elements of the egg, you can not expect the machine to turn out eggs.

Therefore, the scientists analyzed the egg and not only that, but also analyzed the body of the hen with the feathers and discovered as follows: The very large number of different substances found in the hen may be grouped under four heads: 1, water; 2, ash or mineral matter; 3, protein (or nitrogenous matter); 4, fat. The proportion of each of these groups alter with the condition of the hen. Water is the largest ingredient and amounts to from forty to sixty per cent of the weight of the bird. Ash or mineral matter forms from three to six per cent when the hen is not laying, and from six to ten per cent when laying. The groups called protein constitute from fifteen to thirty per cent of the weight. Fat seldom falls below six or rises above thirty per cent.

The feathers are composed of protein and ash, the ash being largely silicate of potash and lime.

The analysis of a fresh egg shows: 1, water 67; 2, ash 12.2; 3, protein 12.4; and 4, fat 8.7. The shell is earthy matter, nearly all phosphate of lime. The white is nearly all albumen or protein and water, and the yolk protein, fat and water.

It is interesting to compare the analysis of the hen and egg with some of our grains and poultry foods, but it would take more time than is permissible in a short talk. In all our grains are found more or less the elements of the egg, but they are not in the right or proper proportion for making the egg. There is usually too much of the fattening element in the grains and not enough protein or nitrogeneous element, which forms the meat, muscle, bone and feather. This is the most valuable and most expensive part of the ration.

In order to keep up the strength of the hen and have her produce the largest amount of eggs it has been found that for every pound of protein in the food, she must have four pounds of carbo hydrates. This will vary slightly according to the heat of the weather and the needs of the hen.

I wish I could go more fully into this interesting and important subject but space forbids it. I would urge you to send a postal to the University of California at Berkeley, asking for the Farmer's Bulletin No. 164 on Poultry Feeding. This bulletin, by Professor Jaffa, is one of the most valuable bulletins ever published. It contains the analysis of the different grains, vegetables and meats and of most of the proprietary foods, besides formulas for the best rations.

The Victory Poultry Ranch of Goodacre Bros.

Feeding in all its Phases

Again I will say that the three essentials of egg production, the three essentials of profit in poultry keeping, the three essentials for vigor and health in fowls are, comfort, exercise and proper food.

Let us consider (1) the proper food, (2) the methods of feeding it, and (3) recipes for a few tried balanced rations.

Practical knowledge and skill in feeding can be acquired without the study of science. Feeding fowls for good results is a comparatively simple matter.

Requirements in Feeding

The food which a fowl consumes has three chief functions to perform: (1) To sustain life, promote life, repair waste and produce eggs; (2) to keep the body warm; (3) to furnish strength or energy which is expended in every movement. The fowl is also able to store food, not needed at the time it is eaten, for future use; this store is chiefly in the form of fat which serves as a reserve supply of fuel.

Food Elements

To supply the three functions in the life of a fowl there are three principal food elements: Proteins, carbo-hydrates and fat; all of these are contained in the different grains and foods used for poultry.

(1) Proteins (or pretein) albuminous or nitrogenous matter. Protein is the nourishing matter, the principal tissue former, supplying material for bone, muscle, blood ,feathers, eggs. Its latent energy can also be converted into heat and energy, but it is more costly for such purposes than the non-nitrogenous foods.

(2) Carbo-hydrates, carbonaceous matter, starches and sugar. Carbo-hydrates form the bulk in nearly all foods and are the principal sources of heat and energy.

(3) Fats are found in almost all foods. They furnish heat and energy in addition to the supply from the carbo-hydrates. Fat also enters largely into the composition of the yolk of the egg.

All three food elements are necessary for life. The proper combinations of these three is called the "balanced ration". It is, in other words, a complete ration containing in proper proportions the necessary food elements to promote (1) growth, including egg production, (2) warmth and (3) energy or strength. The needs of a fowl's system are not always the same; it does not always need the different elements to be in the same proportions; the ration properly balanced (or suitable) for a growing chick would be unbalanced (unsuitable) for the mature hen. The food to be a balanced ration must be adapted to the present needs of the fowl.

Methods of Feeding

The question of how to feed and what to feed for the best results in egg-production, is the most difficult problem in poultry keeping, and has for some time been engaging the attention of the various Government Experiment Stations in this and other countries. The two successful systems in use at the present time are the Mash system and the Dry Feed system.

The mash system is one in which a mash is fed once or twice a day. The foundation of the mash is bran, middlings, and corn meal or chops. It is mixed wet, raw, scalded or cooked. The dry feed system is when a dry mash is fed, consisting of the same ingredients as the wet mash. Dry feeding is used by many regularly and is becoming more popular every year.

The advantages of a mash are that by its means the food ration for the whole day can be properly balanced; the nutritive ratio varied and controlled and the waste vegetables and table-leavings utilized to the best advantage.

In mash feeding the errors to be avoided are: Too concentrated a mash with too much meat or fat; too light or bulky, that is,

An Excellent Feed Hopper, Good Both for Young and Old Fowls.

These hoppers are made 8 feet long and the trough is 8 inches wide and 4 inches deep with a projecting strip on top ½-inch to keep the chicks from pulling our the feed. The slate are 3 inches apart.

composed principally of bran or hay; too wet or sloppy mashes or sour or mouldy. Experience has shown that feeding mashes more than once a day has bad effects, producing indigestion in various forms.

The advantages of the dry-feed system are: A saving of labor to the feeder, is lighter to handle and much easier to mix. It can be fed in the morning. The fowls are obliged to eat it slowly; they cannot swallow it in a few minutes. It will not freeze in cold weather nor become sour in hot weather and the fowls will not over-eat with the dry feed.

The chief consideration in dry-feeding is that fowls require about three times as much water to drink as with the wet mash; also unless the dry food is placed in hoppers or fed in boxes at least four inches deep, it is apt to be wasted. The two systems supply the requirement of the fowls in slightly different ways and both are used very successfully.

SAMPLE RATIONS

The rations here given have been tested and proved excellent by some of the most successful poultry breeders in this country.

Ration for Chicks Intended for Breeders

First meal, when chicks are 36 hours old,—rolled or flake breakfast oats, dry; give scattered on sand every three hours, then feed chick food. This is a number of small or broken dry grains which can be bought at the poultry supply houses. The use of hard grain diet like chick feed, develops the digestive organs and keeps them healthy. The chick feed prepared by reliable firms is excellent. For those who prefer to mix their own chick feed, the following is a good recipe: Cracked wheat, 30 pounds; steel-cut or rolled breakfast oats, 30 pounds; finely cracked corn, 15 pounds; millet, rice, pearl barley, rape seed, finely ground beefscraps or granulated milk, dried granulated bone, chick grit, ten pounds; granulated charcoal, 5 pounds. In the chick feeds wheat, oats and corn are the staples, the most necessary part of the ration. Feed at six a. m. chick feed scattered in chaff; 9 a. m. rolled or stell-cut oats; 11 a. m. green lettuce; 1 p. m. chick feed; 3 p. m. green feed, lettuce, clover or potatoes chopped fine; 4:30 p. m. hard boiled eggs (4 for 100 chicks, chopped shell and all, with the same amount of onions and twice the amount of bread crumbs or rolled oats or johnnycake. One fountain of skim milk and one of clean water always before them and renewed three times a day. Very coarse sand and granulated charcoal should be always before them.

Toward the end of the second week mix a little whole wheat, hulled oats and kaffir corn with the chick food, gradually increasing it until at the end of the sixth week they will be eating this entirely.

Ration for Broilers

For the first two weeks use the same feed as given for the breeders. Third week, 6 a. m., chick feed; 9 a. m., mash, 1 part each of bran, cornmeal and rolled oats, and a little salt; mix with skim milk, making a crumbly dry feed in a small dish or trough, taking away all there is left in fifteen minutes; 11 a. m., lettuce or clover; 1 p. m., rolled oats; 3 p. m., chopped raw potatoes; 4:30 p. m., mash same as in the morning. Fourth week, 6 a. m., chick feed; 9 a. m., mash; adding 5 per cent beefscraps or cracklings; 1 p. m., chopped potatoes; 4:30 p. m., mash, same as in the morning. Keep grit and charcoal always before them, with skim milk and pure water. Finish off at six to eight weeks by gradually adding from five to ten per cent of cotton-seed meal and a little molasses with the mash.

Ration for Laying Hens

In order to keep up the strength of the hen and have her produce the largest amount of eggs it has been found that for every pound of protein in the food she must have four pounds of carbo-hydrates. Many instances may be cited in which the rations fed to laying hens differed greatly but have been productive of excellent results provided they contain a sufficient quantity of digestible protein. The following rations have proven successful:

DRY FOOD METHOD. By measure, 2 parts bran, 1 part alfalfa meal, 1 part corn meal, 1 part rolled oats; 1 part beef-scrap or granulated milk, a little pepper and salt. Keep this in a hopper or feed box. At noon green feed, evening grain; wheat, kaffir corn (or cracked corn), hulled oats, equal parts, mixed and scattered in straw or litter in the scratching pen. Fresh water constantly before them. If they run out of water the egg yield will stop.

For One Dozen Hens

Rations for one dozen breeding hens, American class, in confinement, for three days' rotation.

Monday morning—One pint and a half grain, wheat, cracked corn and hulled oats, equal parts mixed and scattered in straw or litter in scratching pen. Noon: Cut clover or lawn clippings. Evening: Mash, 1 pt. heavy bran; 1 qt. ground oats; 1 pt. corn meal; 1-3 of the whole cut clover or alfalfa meal; 1 tablespoon each of salt and pulverized charcoal; ½ pt. beef-scraps.

Tuesday morning—1½ pts. mixed grain, wheat and rolled barley. Noon: green feed, pumpkins or clover; 1 pt. green cut bone. Evening: Mash, 1 pt. cooked vegetables and table scraps, 1 qt. bran, 1 pt. cornmeal, a little salt and pepper.

Wednesday morning—1½ pt. mixed grain; wheat, hulled oats, kaffir corn. Noon: Cabbage or beets. Evening: Mash, 1 pt. peas or beans soaked over night, boiled with a little soda until soft; ½ pt. dried blood, or beef-scraps, 1-3 cut clover. If you cannot get beans cheaply, use potatoes or other vegetables.

Follow the same system the remaining three days.

Sunday, instead of the mash, scald three pints of rolled barley in the morning, cover and leave to steam. Feed in the evening instead of the mash; this makes a pleasant change and saves work for the Sabbath.

The reason for feeding the mash at night is to keep the hens busy scratching all day and so send them to roost with their crops full. There is danger of the American and Asiatic fowls becoming too fat and lazy without exercise if given the mash in the morning.

Fattening Fowls

Fowls to be fattened should be confined in small yards or in coops or crates, especially adapted for feeding. The object in keeping them in confinement is to prevent the forming of muscle and sinew which would occur if allowed to run at liberty.

The crate used for fattening fowls can be four or six feet long. Mine were composed of lath six feet long; the frame of the crate is 6 feet long, 18 inches wide and 18 inches high, divided into six little stalls or compartments. The frame is covered with lath, placed lengthwise on the bottom, back and top the width of one lath apart. The first lath on the bottom should be two inches from the back to allow the droppings to fall through, otherwise they would lodge on the lath at the back. The lath are placed up and down in the front, the spaces between them being two inches wide to enable the chickens to feed from the trough. A "V" shaped trough is made to fit into two notches in cleats in front of each crate. The crate stands 15 inches from the ground; the droppings are received on sand or other absorbent material and removed daily. The coop is large enough to hold 12 or 18 young chicks (2 or 3 in a stall) or six full grown fowls. Fowls are fed three times a day all they will eat in 15 minutes.

See cut of fattening crate.

Formulas for fattening:

(1) Equal parts of bran, cornmeal and oat meal (rolled breakfast oats) mixed with skim milk, fed three times a day.

(2) Buckwheat flour, pulverized oats, cornmeal in equal parts, mixed thin with buttermilk.

(3) Equal parts barley-meal, and oat-meal and a half part of corn-meal, mixed with buttermilk or skim milk.

(4) A favorite French combination is two parts barley meal, one part corn meal, one part buckwheat flour.

A little salt and coarse sand should be added to their food. Three weeks is the length of time to continue the feeding. Chickens do not seem to be able to stand the confinement for a greater length of time. The last week of the fattening process, five per cent of cotton seed meal and a little tallow may be added to any of the above formulas.

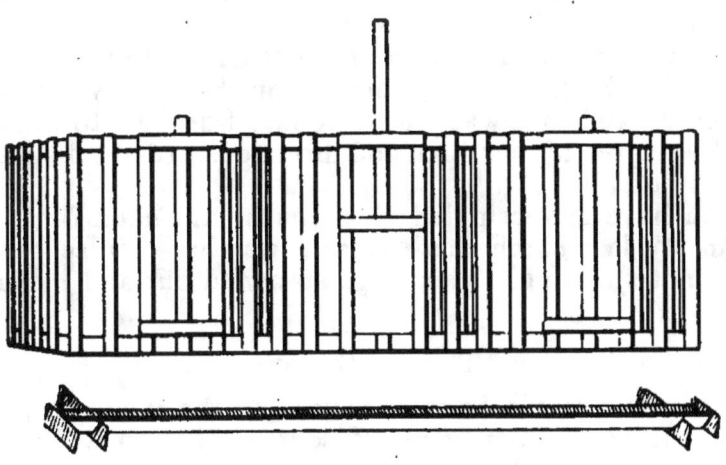

Three-Compartment Fattening Crate.

Testing Eggs for Incubation

Success is what we all want to attain in what ever we undertake, and I earnestly hope that my practical talks on poultry may help others to make a success of it.

"Success with the Japanese," wrote George Kennan, in one of his interesting articles during the war, "is not a matter of perhaps or somehow or other, nor does it depend upon the grace of a merciful God. It is carefully 'pre-arranged' by an intelligent forethought, a perfect system and an attention to details that I have never seen surpassed."

Success in the poultry yard can be attained or "pre-arranged" in exactly the same manner. Failure in the chicken business (as in warfare) is due to lack of forethought, lack of system, and carelessness with regard to details. Forethouht, is the studying up and thinking how to do a thing, thinking out beforehand the best way of doing it and arranging for it.

The experiences of others by teaching us may save us not only dollars and cents but chagrin and disappointment. I spend a good deal of my time in visiting the ranches of some of my correspondents, either to help them out of difficulties, or to mate up their pens for them, or to start up their incubators, or to overhaul their brooders or plan their henneries, and in this way I become acquainted with the needs and difficulties of a number of amateurs or beginners in the poultry business. Some of the troubles of others may teach us what "not to do."

"I wish you could tell me what is the matter," wrote one. "I had good luck last year but only half the fertile eggs hatched last time."

I answered by spending a day at her ranch. "What is the matter with your hatches?" said I, "and on what day did they come out?"

"The first hatch this season came out on the twenty-second day," was her reply, "and as it was a day too late I decided to run the machine half a degree higher than the directions order, and I suppose I got it too hot."

"Did you have any crippled chickens in the hatch?"

"Yes, in the last hatch there were a number of nice big chicks that could not stand up. Their legs sprawled out and I had to kill them."

The Incubator

Cripples usually come from over-heating the incubator, or from irregularity of heat. Poor or insufficient ventilation will also cause cripples.

Now, what was the reason for these failures and what can others learn from them? After a careful examination of the incubator, which was a good one of the most approved make, I decided first that the incubator did not stand perfectly level, secondly that the thermometer was at fault. When the incubator is in the least degree out of level the heat will go to the highest side, leaving the lowest possibly a degree or more too cold. The first thing to be learned from this lady's failure is never to start the incubator without being absolutely certain that it is perfectly level. The only way to do this is to use a carpenter's spirit level. Put it on top of the machine at each side and then cross-wise, and be sure that the bubble of air is at the proper spot. You may think that because it stood level last year it is most likely to be all right this year. That is leaving it to chance. One of the legs may have shrunk ever so little from the dry weather or swollen from the dampness of the room or the floor or ground may have changed ever so little at one corner or side without it being perceptible to the eye. It is much "better to be sure than sorry" so whether you are an expert or not, do not commence this season to hatch without testing your machine with a spirit level. Do not trust to luck—"pre-arrange" and success will be yours.

Test the Thermometer

Do not start the incubator this season without testing also the thermometer. Some friends of mine once bought a new incubator of standard make. The thermometer was guaranteed correct; two years seasoned. They had just received from Canada twenty dollars worth of very choice eggs, and as they wanted to be sure of a good hatch from those prize eggs, they bought this new incubator, although they had a good one. Not an egg hatched! They afterwards discovered that the guaranteed thermometer was two degrees wrong. Do not trust to last year's testing. Thermometers vary, and it takes at least two years to season them.

It is not difficult to test a thermometer, but to do so you must have one perfectly correct and accurate. This you can either borrow from the doctor or from your druggist, or you can take one of your thermometers to the druggist and ask him to test it for you. Then having one that is accuate, take a bucket holding about two quarts of water, put warm water heated to about 105 degrees into the bucket and put your thermometers into it with the bulbs all at the same level. Keep the water well stirred, so the heat will be the same all over. Hold the thermometers in it for fifteen minutes, then read them and note the difference. If your thermometer is half a degree too low, mark on the incubator "Thermometer half degree too low; run incubator half degree lower than directed," or opposite, if the thermometer reads too high. If you buy a new thermometer, after testing it be sure to hang or place it in the correct position. The bulb must be on exactly the same level as the former thermometer, which belonged to the machine. A little difference in height or in the position of the bulb of the thermometer may make a great difference in the heat on the egg tray. You cannot be too careful and particular about these small items. "Pre-arrangement" of these means success.

How to Test the Eggs

After supper when it was dark, we put the trays of beautiful fresh eggs on the dining room table, put the egg tester on the lamp, and then looked at each egg through the tester. Eggs were rejected that were chalky to the touch, or those that had light spots in them or freckled all over with clear places, or thin on the little end, or cracked, or crooked, or in any way misshaped. A few doubtful I left in, marking them "d" (these I subsequently heard did not hatch). It is much easier to detect the imperfect or unhatchable eggs by looking at them with the tester than by merely feeling them. It may be a little more trouble at the commencement but is a saving in labor all through the period of incubation and a lessening in the expense of oil; besides giving more room for fertile eggs and more chance of a good hatch, as the infertile eggs chill their fertile neighbors and draw from their vitality. Therefore do not put eggs into the incubator, or under hens, without carefully selecting them. Poultry keeping is made up of little things, and can so easily be ruined by little things that I will add a word of warning. Do not hold the egg when testing it so close to the lamp that it will heat it. The tiny germ of life in the egg is very tender and may easily be killed. For this reason I made a home-made tester out of a cracker box. I cut a hole the size of half a dollar just opposite the place where the flame of the lamp came when I set it inside the box. In this way I did not overheat the egg. I also found this box very handy for testing eggs under setting hens. Eggs, whether under hens or in incubators, should always be tested out. There are thousands of eggs lost or wasted every year from carelessness in this matter. An egg which is infertile and is for a week either in an incubator or under a hen is perfectly good for food. It is simply an egg that has been in a warm place for a week. There is no germ in it; there never has been life in it so there is no dead germ to decay. Infertile eggs keep fresh and sweet much longer than fertile eggs and those who are raising only eggs for market should keep no male birds in their flock and never have fertile eggs.

Do not put eggs from different classes of fowls into the same incubator. Hens' eggs take twenty-one days to incubate, but if eggs from Leghorns (Mediterranean class) are placed in the same tray with Brahmas (Asiatic class) or with Plymouth Rocks (American class) the Leghorns will be the first to hatch, sometimes as much as two days earlier, to the great detriment of the larger breed, which is slower in hatching. This comes not only from the earlier hatched chicks walking over the eggs, but also from the change in the atmosphere and temperature in the incubator at the time of hatching. At that time the air in the incubator is always heavily charged with moisture and the temperature rises from the activity of the chicks, and these two conditions will ruin the hatch of the slower breed. Experiments along these line that I have made have always given the same results.

Natural Incubation

The beginner may find it best to incubate with hens in preference to an incubator. The hen, having layed the egg, is the natural mother, has the mother instinct given by the Creator, and is certainly the one intended to hatch and brood the chickens. To the beginner in the chicken business there is less present outlay in a few setting hens than in installing even a small incubating and brooding plant under artificial methods. The trials of those who find setting hens troublesome are mostly due to their own inability, or their lack of patience with the hen. Hens must be treated with patience and gentleness, for in no way can a hen that has the "setting fever," as our grandmothers called it, be coerced against her will.

How to Make Nests

The nest should be about fourteen inches square. Some breeders use boxes twelve by sixteen inches, but I prefer the square nests. If the nest is to be on an earth floor, rake the floor, then scoop a place about thirteen inches across in the form of a saucer; firm the shape well with the hand, and when it is smooth and firm, take hay or short straw, or tobacco stems and firm that again in the proper shape, and the nest is made. Should it be necessary to have the nest in a box or on a board floor, take a clean box, have the front of the box just high enough to retain the nesting material; the backs and sides may be higher; put several inches of fresh earth into the

Nature's Way of Hatching.

box, firm it with the hand into a saucer-shaped hollow, and be
sure to pack the earth high into the corners, so there will be no possibility of the eggs rolling into a corner and being chilled or lost.
The nests should be flat at the bottom, shaped like a saucer and
not like a bowl. If too deep the eggs will roll together, sometimes
pile up and get cracked or broken.

When only a few hens are to be set, the nests can be placed in
any convenient location where the hens may be quiet, comfortable,
away from other fowls and in the shade. I have found that trap
nests with two compartments very satisfactory, placed under a tree.
I also have made sets of nests, giving each hen a nest and a small
run, with a dish of water, a hopper with grit, corn and wheat always
before her, shut off from all intruders. If hens are to be set in large
numbers, a separate hennery in which from six to twenty hens can
be set on the same day is the most convenient. The nests in this
house or room should be placed with their backs to the wall and
should face towards the center. Grit, corn, water and a dust bath
for them to bathe in must be before them at all times. After a few
days, if this hennery has a separate yard from the other fowls, the
door of the house may be left open so the hens can go out of doors
and take a dust bath in the open air, but the food, water and grit
must be in the house in sight of all the hens.

Setting the Hen

The old fashioned recipe was, "Set a hen between sunset and sun-
rise" for luck. In other words, set a hen in the dark. Hens are
quieter and not so easily frightened after dark. Choose quiet, gentle, tame hens; they make the best mothers. Handle them very
gently. Put all the hens on the eggs in the same room the same
evening, so they may all hatch out the same time. This is in order
to keep the hens quiet during the hatch, as some whose eggs were
not hatching the same day might become so excited they would
leave their own nests and try to get to the newly hatched chicks
when they heard the first peep.

Dummy eggs should be placed under the hens, when a number of
hens are set in the same room, for a few days, a few under each
hen. The first night after dark set all the hens on dummy eggs. If
some light is necessary, turn the dark side of the lantern toward
hen. Have as dim a light as possible; move the hens gently. They
will soon settle down on the eggs. In the morning look in and if
any hen appears refractory put her on the nest again and cover her
with a box. Look in frequently for the first few days to see how
they are doing, and you will rarely find more than two hens off and
eating at the same time, as they are afraid of leaving their nests
when others are off. Let the hens sit for two or three days, then
put the good eggs gently in at night. The way to do this is to remove the hen gently, setting her on the floor, take out the dummy
eggs and put the real eggs into the nest and gently replace the hen.
Do not talk, act quickly, silently and swiftly, in a very dim light.

From thirteen to fifteen eggs are all that should be placed under
a hen. It is all she can warm properly, all she can turn and attend

to without the risk of breaking or cracking some. You will hatch more and stronger chicks by not placing too many under a hen.

Keeping Records

Above each nest, hanging on a nail, I place a card. On this card, legibly written is: (1) The date when set; (2) when due; (3) the hen's name or number; (4) name or parents' number on eggs; (5) number of eggs; (6) date of first test, number infertile or dead; (7) date of second test and remarks; (8) hatch, number taken from nest, number not hatching or killed; (9) toe marks of chicks. These cards can be preserved or copied into the diary of the ranch. They form a complete data of each hatch and a history of the hens as well as the chicks.

Testing the Eggs

Watch the hens rather closely for the first week, and note any that may be restless, nervous, cross to the others or stupid in not finding their way back to their own nests. These, when you test the eggs, you may be able to cull out and turn them back into the laying pen. It is always best to keep hens of pleasant disposition for mothers.

The eggs should be tested about the seventh day. An expert can test them earlier, and white eggs or duck eggs show the germ as early as the fourth or fifth day. The removal of the infertile eggs gives those that are left a better chance of hatching. The infertile eggs or dead germs are colder than the living eggs and chill the latter; besides, the infertile egg has a market value and can be used in the kitchen or fed to the chicks. It is a waste to throw them away. Testing should not be neglected. There is no use in hens setting on eggs that will not hatch. They had better be reset on fresh eggs or returned to the laying pen.

Egg testers can be bought at the poultry supply houses, but a home-made egg tester I have used for years is only a box with the back knocked out and a hole in the top for ventilation. I put the lantern into it. Just opposite to the flame a hole about two inches square is cut in the box and a piece of a rubber boot leg tacked on. I drew a pencil line around a fifty-cent piece and cut that out with a pen knife, leaving the round hole for the light to shine through.

The testing must be done in the dark. Set the egg tester with the lantern inside it on a box near the nest. Take the hen quietly off the nest, being careful to put your hands under her wings to make sure that you do not lift an egg or two with her. Place the hen very gently on the floor at one side. Do this so gently that the hen will not realize that she is off the nest. Take all the eggs from the nest, placing them either on the floor or in a basket; examine each egg and replace each fertile egg in the nest as you examine it; mark on the record card the number of infertile eggs, and gently replace the hen on the nest. Should any hen awake and appear nervous she can be put upon the nest and the eggs slipped one at a time under her as they are tested, but the former plan is preferable, being more quickly done, with less disturbance to the hen.

The light shining through the egg, when held against the hole in the tester, shows the condition of the egg. Infertile eggs are clear. Fertile eggs have a shadow in them by the seventh day. The germ appears in some like a dark, irregular floating spot. Doubtful eggs should be marked with a D and given the benefit of the doubt, replacing them in the nest.

After taking out the infertile eggs, if there are many of them, you can reset the hens that have none or turn them back into the laying pen, culling out the fractious or nervous hens. By doing this carefully at each test you will probably have good mothers when hatching time comes. Restless setters usually make indifferent mothers. Close observation is necessary for success in all lines of poultry culture, and especially with setting hens.

The second test should be made in the same way on the fourteenth day. The eggs containing dead germs should be buried.

Dusting the Hen

A hen should be well dusted with insecticide the day she is set. To dust a hen the powder should be in a tin box with a perforated cover. An effective home-made peppering box can be made from a baking powder can with holes in the lid. Hold the hen by the legs, lay her on her side on a newspaper, raise the wing and sprinkle under it, then rub the powder well into the skin, especially round the vent. Work it into the soft feathers also around the neck. When one side is thoroughly powdered turn the hen over and do the other side. The powder that is spilled on the paper can be returned to the can.

While the hens are on the nests they should be dusted on the seventh and fourteenth day and two days before the hatch comes off, with buhach or with any good insecticide. I prefer those principally made with tobacco dust.

When Hatching

In the climate of California I have never found it necessary to moisten hens' eggs. In fact, the eggs that contain dead chicks show that they have not dried out enough. They did not require more moisture. There is a natural perspiration which comes from the hen, and this keeps the eggs moist enough.

Should the eggs be chilled by the hen deserting the nest, do not throw them away. Put them under another hen as quickly as possible. I have known of eggs being left for a whole day and yet hatching. Eggs under hens will stand much more cooling than in and incubator. Chilling seems to be less injurious during the second week of incubation than at any other time.

On the nineteenth day, two days before the hatch, I take out to the nest a bucket of warm water, temperature 103 degrees; removing the hen from the nest I put the eggs into the water. Those with a live chick in them immediately begin to bob or move as they float on the water, and I return them to the nest; those that sink to the bottom or remain perfectly quiet have dead chicks in them and will not hatch, and I mark them with a pencil; then replace the hen upon the damp eggs, feeling sure I will have a good hatch.

It is best to watch the hens pretty closely when the chicks are hatching. Some hens get excited and nervous when they hear the chicks peeping, and in their restlessness crush the shell so that the chicks cannot turn themselves and they die in the shell. These nervous hens should, if possible, be removed and quieter hens put on.

When chicks are hatching rapidly and the hens are nervous, it is best to remove the chicks as they dry off, taking them to the kitchen in a basket lined and covered with flannel. But if the hens are quiet it is best to leave the chicks with the mothers, only visiting the nests about twice during the hatch to take out the empty shells, lest they should slip over the yet unhatched eggs and so smother the chick. All eggs should be hatched by the end of the twenty-first day.

Marking Chicks

The offspring of the best, or pedigreed stock, can be marked so as to know them through life by having a small hole punched in one or more of the webs of the feet. This should be done as the chicks are removed from the nests. A marker or punch is sold at poultry supply houses for marking chicks. They should be marked the day they are hatched, as the web is then soft, does not bleed as much as later, and there is not as much risk of the other chicks pecking the toes as they would do when older.

If the hens have been well cared for, properly dusted with a good insecticide during the three weeks of incubation, they will be perfectly free of lice. They and the chicks must be kept free. There is not the difficulty in this that many imagine. Dusting the chickens and hens once a week is is all that is necessary. Some breeders put a little lard on the top of their heads and on their throats. This protects from the head lice. Others take a small brush ·(if the chicks are affected with head lice), and wash the little heads once a week with a lather of carbolic soap. They soon dry off in the sun or under the hen.

White Leghorn Cock.

Artificial Incubation

We are living in wonderful times, in the age of great inventions, and to succeed in any business, we must keep abreast if not ahead of our times. Not the least wonderful accomplishment of this wonder-working epoch has been the growth and advancement of the poultry industry, and the invention of the modern incubator, which made the development of the poultry business in this country possible.

In Egypt and China artificial incubation has been known and practiced for many centuries. In this country it is scarcely out of its infancy, still it would be impossible to estimate the value of the incubator to the poultry industry. It has made possible and profitable the large poultry plants in this country. It has developed the broiler business; it has raised the hen to the position of the money maker. One incubator will do the work of ten to thirty hens and with better results.

Must Approach Nature

There have been many kinds of incubators invented, made and patented in the last twenty years. The difficulty is to choose which kind will do the work of hatching eggs best: that is, will bring out strong chicks with the least attention and the least expense . There are hot water machines and hot air machines; round incubators and square incubators. I have heard of incubators in this State, which are made like hot beds heated with stable manure. Some incubators are heated with gas, but most of them by the heat of a lamp which burns coal-oil. The best incubator is the one that comes nearest to imitating the natural process of incubation by a hen, for undoubtedly Nature is our great teacher in this matter.

The two favorite makes of incubators on the market now are the hot water incubators and the incubators which bring warmed air into the egg chamber. The latter are called hot-air incubators. The difference between them is that the hot water machines heat the egg chambers by radiation, while the hot-air machine brings warm air into the incubator.

In the machines where the heat is radiated from the metal surface of pipes or tanks the temperature at the underside of the eggs, away from the heat, is several degrees cooler than at the upper side of the eggs. Top heat by radiation is supposed to resemble the heat from the body of the hen.

In the hot-air incubators, the egg chamber is heated by air that is warmed outside of the egg chamber to a proper heat and is then forced into the machines by suction or circlation and diffused into the egg chamber. This way gives a constant supply of warmed fresh air, as pure and fresh as the atmosphere outside of the incubator. These hot-air machines rarely require any moisture to be added, as there is usually sufficient moisture held in suspension in the atmosphere, which is being constantly introduced into the egg chamber.

It pays to get the best, and by inquiring at the large poultry plants in the neighborhood, information can easily be obtained as to the most popular machine in use in that locality.

It is wiser to buy a machine than to attempt to make one. Good incubators are now sold at so low a price that it does not pay to risk the loss of eggs in experimenting on a home-made machine.

Location of Incubator

The incubator should be located in a well ventilated room or cellar that is dry and not subject to great variations of temperature.

Preparing to Hatch

The first thing to do is to set the machine perfectly level, using a spirit level to make sure of this, for if the machine is not level the heat will go to the higher side, the temperature will be uneven and although it may be correct where the thermometer hangs, in the middle, the upper side will be too hot and the lower too cold. It is most important to have the incubator stand perfectly level.

Let the incubator run for thirty-six hours before putting in the eggs. This is to make sure that the machine is thoroughly warmed and that it is running steadily at the proper heat. It may take twelve hours before the eggs gradually warm through, and the thermometer again shows the desired temperature. During this time the regulator must not be altered. Touching the screw may prove fatal to the whole hatch. So wait patiently until the desired heat is again present.

Selecting the Eggs

Eggs for hatching should always be carefully selected. The fresher they are the better. Eggs hatch after being kept a month, but the little germ or seed of life gradually grows weaker and weaker, and at last has not the strength to develop into a fine, healthy chick, and may die in the shell, if the egg is kept too long. Ten days or two weeks is better than any older.

The eggs should come from vigorous, healthy and well-fed stock. Much depends upon the feeding of the breeders, especially the male bird. They should have plenty of vegetables and green food, as well as animal food and those grains which contain the bone and muscle forming elements. Eggs with imperfect shells should be rejected; also those with rough or chalky shells, and with thin spots. The eggs should be of medium size, neither too large nor too small, as the large eggs may have double yolks, which rarely hatch. Small eggs denote inferiority and are either pullet eggs or eggs from fat hens, or hens exhausted from having layed a long time.

Eggs of One Class

The eggs should be of one breed or class. It takes twenty-one days to hatch all hen eggs, but if the eggs from Leghorns are placed in the same tray as the Brahmas, the Leghorns will be the first hatched, sometimes as much as two days sooner, to the great detriment and loss of the others, which are slower in hatching. This is probably caused by the change in the atmosphere and change in the incubator at the time of hatching. The air is heavily charged with moisture, and the temperature always rises during a hatch from the activity of the chicks, and it is exceedingly difficult to regulate the temperature when the incubator is full of chicks in all stages of hatching. The rise of temperature does not hurt the chicks that are just breaking out of the shell, but if it takes place two days too soon it will ruin the hatch of the heavier and slower breeds. Experiments that I have made along these lines have always given the same results.

Turning the Eggs

The eggs must be left for forty-eight hours after being placed in the incubator before being turned. After that, they should be turned twice a day, or oftener. In this we should imitate the hen, for she not only turns her eggs constantly, but always shifts their position, pushing those that are on the outside into the center of the nest. It is really more important that the eggs be moved or shifted from their position or location in the tray, than merely turned, as it shifts the locations of the eggs in regard to weak germs or infertile eggs.

If the eggs are not turned during the early stages of incubation, many of the germs will dry fast to the shell and die, and the egg will be lost. When the egg is not turned during the latter part of incubation, the embryo does not develop properly, has little chance of hatching or may prove a cripple.

The turning and moving of the eggs gives exercise to the embryo; it is a species of gymnastics for strengthening the chick. The first forty-eight hours and the last forty-eight hours the eggs must not be turned.

Cooling the Eggs

Cooling the eggs I consider an important matter in our American incubators. The first week, following the hen's example, the eggs require but little cooling beyond the time it takes to turn them. The second week, as soon as the eggs are turned replace them in the machine and leave the door open for five minutes, after this increase the time, a minute or two each day till at the end the eggs are being aired or cooled fifteen or twenty minutes.

Cooling the eggs helps to make the shell brittle, so that the chick at the proper time can break its way out. Cooling the eggs contracts the shell and heating it up again expands it and this contraction and expansion gives the shell its proper brittleness. As the eggs warm up again, an almost imperceptible moisture comes over them, which takes the place of the perspiration of the hen, and obviates the necessity of sprinkling or dampening the eggs. So in our incubators it is necessary to cool the eggs. If this has been done properly the chicks will be strong and vigorous and few will die in the shell.

Testing the Eggs

All sterile eggs and dead germs should be tested out. Egg testers are sold with all incubators and very little practice will enable even a beginner to detect the sterile eggs and dead germs. Infertile eggs will be of a clear, uniform color throughout, except a slight darkening where the yolk lies. In the fertile eggs will be seen a small dark spot, and in a white egg the blood vessels can be seen branching out from it. Eggs should be tested about the seventh day. A second test for removing the dead germs should be made on the fifteenth day, they being easily detected at that time. The chicks in fertile eggs will be seen to fill the shell nearly, except a small space at the small end, and the air space at the large end. All eggs containing dead germs should be removed from the machine and buried. On the eighteenth day the chicks fill the entire shell except the air cell and the egg will be quite opaque, as if nearly full of ink. To become accurate in egg testing requires practice and a brilliant light.

Operating the Incubator

Follow exactly the directions given with whatever incubator you may purchase. The makers of the incubators are anxious for you to succeed and have good hatches; it is to their interest for you to be successful. They have spent time and money in perfecting and understand how to manage their own machines better than any one else.

On the morning of the nineteenth day the eggs should be turned for the last time. The machine should then be closed and kept closed until the hatch is over. Opening the door during the process of hatching may spoil or seriously injure the hatch, as by such action a large amount of heat and moisture escapes and cold air is admitted. This dries up the lining skin of the eggs that are pipped and checks or prevents their hatching. It also chills the half-hatched or newly hatched chicks and is detrimental to all of them. When the chicks are coming out lively the temperature will rise; should it go above 105 degrees the lamp may be turned down a little.

Leave the chicks in the machine without opening it until they are thoroughly dry. The chicks should not be moved from the incubator until the twenty-second day and should not be fed until twenty-four hours after hatching.

General Remarks

Should the hatch not come off until after the twenty-first day, it shows that the heat has been insufficient; if it comes off earlier the heat during part of the time has been too high. Too low a temperature will give a weak hatch, many chickens will die in the shell, and those that are hatched will be weakly and never amount to anything. Too high temperature at the commencement of incubation will cook and kill the germ. One hundred and six degrees is danger point up to the tenth day. Germs which died between the first and second testing are frequently the result of overheating. Too high a temperature during the last week will so weaken the bowels of the chicks that they will be unable to assimilate the yolk of the egg. The yolk of the egg is Nature's perfect nourishment, which feeds and nourishes the embryo.

During the last day of the chick's life in the shell the part of the yolk which has not been absorbed is drawn up into the chick. This forms its food and nourishment for about three days. But should the egg be over-heated, this yolk hardens and even if drawn into the chick, it becomes tough, the chicken's bowels are weakened by the over-heating, the yolk remains unassimilated, like a piece of rubber, blood poisoning ensues and the chick dies some time between the first and tenth day of its life. Chilling the eggs has almost the same effect; it weakens the bowels, hardens the yolk and eventually kills the chick.

Weaning Coop for Chicks.

Care of Brooder Chicks

The hatching of chicks is but half the battle, for eggs from good vigorous parents will hatch with but little trouble if a good standard incubator is used and if the directions with it are followed. How about the raising of the chicks after they are hatched?

The poultry papers agree that there is not a subject pertaining to poultry culture that needs more thorough, painstaking investigation and discussion than the care of the chicks, and it is said that not more than fifty per cent of the chicks that are hatched the country over, reach maturity or a marketable age.

What are the principal causes of mortality among chicks; how can we combat them and what are the essentials in the successful raising of chicks?

There are numberless causes for the death we deplore—among these are diarrhoea, bowel trouble, lice, improper feeding, impure water, over heating or chilling and exposure to the elements.

Feeling sure that the mortality in chicks is caused in a majority of cases by the carelessness or ignorance of the care taker, let us discuss this subject and glean from the best authorities some ideas about it as far as we may in one short article.

Expert Opinion

Prof. James E. Rice, of Cornell University, has for several years been making a careful study of the cause and cure—or prevention —of the numerous diseases that cause the death of hundreds of thousands of chicks yearly, and his investigations have led him to believe that one great cause of mortality is the failure on the part of the digestive organs of the chicks to properly digest the yolk of the egg remaining in their bodies at the time of hatching.

Mr. Rice says: "If we can solve this one problem—the cause of the anaemic condition of chicks that follows this failure to absorb the yolk of the egg—more money will be saved in one year to the

A Bunch of Future Prize Winners.

farmers and poultry raisers of New York state than it costs to run the State Agricultural College for ten years."

Mr. Rice says he is confident that environment has little, if anything, to do with the disease, as has been generally supposed. When he first began his investigations, this theory was worked upon and followed up, but as the investigation progressed it was found that the same conditions existed under almost any and all circumstances —in dry places, in damp places, in light brooding houses and in dark brooding houses; in fact he found no conditions under which this trouble did not exist. Mr. Rice is confident, however, that the investigations being conducted will ultimately solve the problem.

Until this problem is solved we shall have to be content with the theories of the different breeders and hatchers, and as one I feel confident from my own experiments and experiences that the deaths from diarrhoea, or in fact almost all the deaths of brooder chicks before three weeks of age, come from faulty incubation. The temperature has been either too hot or too cold, usually the former, or the ventilation has been at fault, or the chicks have been chilled in carrying them to the brooder, or fed too soon, before the digestive organs were ready to digest the food.

Elbow Room Needed

Mr. Hunter, the veteran poultry man says: "With incubator chicks raised in brooders, elbow room seems to be a most important factor, and want of elbow room is one cause of the great mortality in brooder chicks."

It is quite natural to suppose that a brooder which is three feet square is abundant room for seventy-five or a hundred chicks, and indeed it is for the chicks as they come out of the incubator, and if we do not want them to grow it might be all right to crowd them into the brooder, but these chicks will be almost twice as large at three weeks old as when they are hatched and will require twice as much room or will suffer for it.

Fifty chickens are as many as should be put into any brooder. To increase the number beyond that point will induce crowding, which kills some and stunts others, and will prevent the quick, healthy growth that is necessary for all young animals. Ample brooder room is the first and chief requisite for the health and comfort of the chicks. The next requisite is oxygen. In other words, plenty of fresh, warm air, but no drafts in the brooder. Here is one of the great faults with many brooders, as for example the hot water pipe brooders in use in many brooder houses. Those hot water pipes merely heat the air that is already within the hovers, which air is practically confined to the hovers by the felt curtain in front, provided to keep in the heat. It does that, but it also encloses the air, which the chicks have to breathe over and over again. This defect in my brooders cost me the lives of many chicks before I discovered the cause. A current of warmed fresh air supplied under the hovers overcame this difficulty, when I substituted the hot air plan.

Comfort Essential

The brooder should be heated for at least twelve hours before the chicks are put into it. I always keep a thermometer in the brooder, and have it at 95 degrees when they are first removed from the incubator. They should be carried to the brooder in a basket lined and covered with flannel, great care being taken that they be not chilled on the way. I am sure that many chicks lose their lives by being chilled on this their first journey. The abrupt change from the warm incubator to the outside air, which is thirty or forty degrees colder, is sufficient to chill the chick.

A chill will harden the yolk of the egg, which is drawn up into the chick the last day of its stay in the egg shell. You know that the yolk of the egg forms the nourishment for the chick inside the shell. The last day of its life in the shell all that remains of the yolk, about one-fourth of it, is drawn up into the chicken through the navel. If the chick is vigorous the yolk should be assimilated or digested in about three days. But if the chick is chilled or overheated, it so weakens the the bowels that they cannot digest the yolk or absorb it, and the yolk hardens or toughens, becomes almost like rubber; then it can never be assimilated, blood poisoning ensues and the chick's life ends.

Chicks should not be fed for from thirty-six to forty-eight hours after they come out of the shell, because, first, they do not require any food, as the yolk inside them takes nearly three days to become absorbed or digested; and, secondly, if they are fed too soon (that is, before the yolk is digested), the effort of digesting the new food draws the nervous energy or gastric juices away from the part containing the yolk, up to the crop and gizzard, and the yolk either does not digest at all or digests so slowly that it brings on bowel trouble, which at such an early age stunts the growth, if it does not kill the chick. In a chick that is fed too early in life the yolk will take, or may take ten days to digest. You ask how I know this. "By sad experience and post mortem examinations," is my reply.

The brooder being warmed to a temperature of 95 degrees under the hover, the floor should be covered with coarse sharp sand, the chicks carried carefully to the brooder, after remaining thirty-six to forty-eight hours in the incubator.

Feed Carefully

The first few hours in the brooder they require no food but the sand to eat and water to drink. The sand supplies the little gizzards with the necessary teeth or little grindstones, so that they are ready to commence work when the food comes. Water I place in a drinking fountain, so they cannot get into it and wet themselves. I give them water from the first. I know some people do not, but it has succeeded well with my chicks. At about four o'clock they have the first meal. I scatter rolled breakfast oats on the sand. The white flakes quickly attract their attention and they pick them up. I also give them a fountain of fresh water and one of sweet skimmed milk. It is surprising to see how quickly they learn to eat and drink. In the evening I look in upon them and am pleased when I see them spread over the hover floor, as it indicates that they are comfortably warm and will not crowd or huddle during the night. The first thing in the morning I give them some more rolled oats and some "chick feed". The "chick feed" I buy at the poultry supply stores. It is composed of a variety of seeds or grains, with a little charcoal, dried blood, or beef scraps and grit. Sometimes I make my own chick feed by mixing cracked wheat, kaffir corn, millet, steel cut oats, pearl barley and rolled oats together, adding charcoal and dried beef scraps. I put more wheat and more oats into this mixture than any of the other grains. The chick feed that I buy has in addition some other seeds, such as rape or mustard, canary seed, hemp, etc. I buy chick feed to save myself the trouble of mixing. Chick feed and rolled oats is their main feed until they are six or eight weeks of age. I feed them five times a day at first, and I always leave a little feed trough or hopper of chick feed where they can get it. I know this is contrary to the advice of many, but I found the weaker ones did not get the proper amount when all rushed for the food, and also it was a great comfort to me, if anything detained me beyond the usual feeding time, to know they had food before them. Also when fed at the usual hour they were not so ravenously hungry; they would not overload their little stomachs.

Their morning meal at about six in the morning, consists of rolled or flake breakfast oats, next green feed, then chick feed, then rolled oats, green feed, and the last feed after they are a few days old is hard boiled eggs (two for every fifty chicks), chopped fine, shell and all, mixed with dry bread crumbs or cracker crumbs, and an onion chopped very fine. I mix all together, adding a little pepper and salt. If I have no bread crumbs, I add johnny cake or rolled oats to the onion and eggs. I always send them to bed with their little crops full.

As They Grow Older

I keep a thermometer under the hover in the brooder and lower the temperature one degree a day until it is down to sixty-five degrees. After the chicks are six weeks old, unless the weather is unusually cold, they require no heat. For green feed they seem to prefer lettuce to anything else. Finely cut clover or alfalfa is excellent. The lettuce I cut up very fine at first, but in a few days they learn to tear it up, and a lettuce suspended on a string or even thrown on the ground, gives them exercise and amusement as well as food.

In the playroom, where the chicks are fed, the floor is covered with chaff. If I cannot get from the mill real chaff I cut up hay in the clover cutter, either wheat hay or alfalfa hay, to give them something to scratch in, and I throw a handful of chick feed into it for them to have something to reward their efforts.

The alfalfa hay or chaff keeps them busy and exercising and this broadens their backs and increases the size and vigor of the egg making organs which are already commencing to grow and which we must develop from the very first if we want to increase the egg output. The chaff, or preferably the alfalfa hay chopped short, also conceals their little feet from their active and sometimes mischievous brothers and stops them from pecking the feet and drawing blood, which tastes so good that they will actually turn cannibal and tear out and eat the bowels, sometimes causing great loss. This is always prevented by keeping the chicks busy scratching in deep chaff.

They have fresh water each time they are fed. The first meal is at about six in the morning, and if I fear that I may be later than that, I put fresh feed and water in their playroom over night, so that the hungry babies may not be kept waiting. They come out at daybreak, eat a little, and sometimes drink, and then go back and take another nap.

The brooders must be cleaned twice a week the first week, three times a week afterwards, and every day when the chicks grow larger. The chicks should be dusted with insect powder about once a week. To do this I have a tin box (a baking powder can with a perforated cover), put insect powder into it and after dark raise the hover and sprinkle the powder liberally over the chicks. This will usually keep them free from lice.

"WHITE DIARRHOEA" IN BROODER CHICKS

This is a disease which rarely attacks chickens hatched and raised by hens, and therefore it must be caused either by faulty incubators or wrong "mothering."

We all know that at times quite a number of chicks in a brooder will be "stuck up behind" as it is sometimes called; how they run about with their shoulders up, looking wizened and old; how they try to huddle near the warmth and finally give up the hopeless struggle and die.

"I think my chicks are taking some disease and dying from an epidemic," said a lady, who, though a novice with incubators and brooders, was an old and most successful poultry woman with hens. These chicks had been overheated in the incubator I discovered, two days after hatching.

Another friend, a very cleaver surgeon, told me one chilly night his incubator lamp went out and all the eggs got stone cold. His wife could not bear to think of losing all those nice eggs after having watched them for nearly three weeks, so she advised lighting up again in hopes of saving some. This they did and were rewarded with fifty nice, lively chicks, but in a few days they commenced to die; they were "stuck up behind" or they shivered and seemed quite thirsty, and at last, when only fifteen were left, he made some post mortem examinations and he found the yolk of the egg, which is drawn up into the bowel cavity the last day of incubation, was still there, only it looked in some like a bit of rubber, in some like hard-boiled eggs, and again in others it was dark and putrid. Instantly he reasoned that it was that yolk that was killing the chicks by blood poisoning.

He had only fifteen left but he decided to experiment on them, so he opened them; his wife begged him to give them chloroform, which I believe he did, and he removed the toughened yolk, sewed up the wound, fed them lightly and all of the patients recovered and lived to maturity.

It was a delicate operation, but my friend had the skillful hand of a trained surgeon. I never attempted it myself, but have made many a sad post mortem on little chicks dying from being "stuck up behind," for I make it a rule to hold "post mortems" on all subjects that die in my yards.

One time a whole incubator of eggs—240—were overheated by a meddlesome child playing with the regulator. Two days later 117 hatched, the others were cooked hard. Every one of the 117 died although some lived to be eleven days old. I did everything I could think of to save them (except the surgical operation) but lost all.

I feel sure that either overheating or chilling so weakens the bowels that they cannot digest, or, rather, assimilate the egg, and that the yolk putrifies and causes blood poisoning; and that either overheating in the brooder or chilling before the chicks are a week old, will have the same result. Also if the chicks are fed too soon after hatching, the digestive juice or whatever it may be called, goes into the crop and gizzard to digest the new food and the yolk of egg is left to either digest very slowly or to not digest at all. In either case it will give diarrhoea and it may end fatally.

I am often asked what to do for young chickens that have diarrhoea, and also for those that are "stuck up behind." I know how almost hopeless these cases are, as they usually come from the un-assimilated yolk of egg, but I reply that rice boiled in milk, adding a teaspoonful of ground cinnamon to every pint of milk is about the best remedy for diarrhoea that I have tried, and to pick off with the fingers the dried excrement, slightly greasing the vent with carbolated vaseline is the only way for "stuck up." If the droppings are washed off it is almost sure to chill the already weakened bowels and result fatally.

A Good Start.

Summer Work

Summer is our time for rest from hatching and now our energies must be directed to safely carrying through the summer the booder chicks and helping the older hens to shed their old clothes and come out in fine and glossy raiment as expeditiously as possible.

Let us first look over our youngsters and see how we can keep them growing. They need a motherly and watchful eye and ear, and a watchful nose also, as much as children do.

Our own lives are made up of little things, but a little chick's life is made up of infinitely little things and it is through little things that success is attained or failure courted. "Be sure to keep the pullets growing," was the vague order given in one of the poultry books that years ago I was studying. The author did not tell how to keep them growing nor did he mention what would prevent them growing, and I just hated that man, but since then I decided that, poor fellow, he most likely did not know himself and was only dealing in generalities to write a plausible article for his book or paper without definitely saying anything. But he was right, we must keep the chickens growing and at the first indication that their growth has stopped we must investigate and find out the cause.

What are the chief causes of chickens not doing well in the summer? Lice and mites. If your chickens are not doing well treat them for lice even if you cannot see them and give their house a good spraying with kerosine emulsion and a little carbolic acid.

Comfort and proper food are the two great factors that will promote the growth of our chicks, and cleanliness is the first requirement. The drinking vessels at this season of the year require special care; whatever may be used should be kept scrupulously clean. I find a sink brush is an excellent thing for scrubbing out the drinking vessels. They must be kept in the shade. They can be placed in a box set on its side or under a shed or tree, and besides being shaded they should be frequently replenished during the day.

Sunshine and Shade

Provide shade for the growing chicks; shade from the burning rays of the sun. Nothing is more conducive to health than sunshine but it must be tempered by shade. Trees and bushes supply the best shade as the temperature close under growing green leaves is several degrees cooler than under anything that is dry or dead. Few realize what a necessity shade is to fowls.

If an epidemic siezes the half grown chicks, it is attributed to any cause on earth but the lack of shade, when in very many cases this is the sole cause. Vertigo, blindness, stunted growth may all be due to the glare of the sun on unsheltered yards. Shade is a necessity and if trees or shrubs are lacking a good shelter can be made by driving a few stakes or small posts into the ground and making a frame upon which palm branches or brush can be laid. I have found a very serviceable temporary shade can be made by ripping open a common gunny sack and nailing four laths on the edges. This little frame can be laid across the top of a small pen or even hung on wire fence and afford a grateful shade.

Overcrowding or the chicks huddling for even one night may stunt the growth or be the means of bringing on an epidemic of colds which may result in roup.

But how to stop them crowding? A mother hen often solves the difficulty by taking the half grown chicks on the perch with her, but for brooder chicks some other plan must be found; the best way is to divide them into flocks or colonies of only twenty-five in each, and supply comfortable perches for them. The chicks will in a short time take to the perches of their own accord.

At one time I had not enough colony coops and a great many chicks. I put them a hundred together in my regular henneries, but they crowded and I not only was losing every night some of the best, but the survivors looked very badly. They sweat off in the night all they had gained during the day. I realized that this meant failure for me if I could not control it. I spent my evenings going around and patiently placing the chicks, hundreds of them, on the perches till I was completely tired out, when I decided to make it so desperately uncomfortable for them they could not crowd.

I bought a bundle of six foot lath and made a lath platform or floor, by nailing them one and a half inches apart, the width of a lath, on stringers one inch by three. This made a flooring of small lath perches three inches above the ground, and make it so uncomfortable for the chicks to crowd that it entirely prevented it. I placed regular perches four or five inches above the lath floor and in a few nights on making my nightly rounds with my lantern I had the satisfaction of finding all the chicks on the regulation perches. I have recommended the lath platform or floor to many and it has proved always successful.

The Proper Range

I would advise you to let the young chicks have free range, and when the pullets begin to show signs of maturing, or at any rate by the beginning of October, to put them into their permanent winter quarters, and to confine them so they will be under your control. They will lay more eggs, if they do not range too far. It has been proved many times and with different breeds, that hens in confinement lay more eggs than those that run at large. The hens can be watched better, are less liable to suffer from maladies; the nests can be kept cleaner and the eggs gathered more easily, while on free range many eggs are lost, nests stolen and the hens will acquire the habit, which we are breeding out of them, of laying only a few eggs and then wanting to set.

In reply to the question of pullets or hens, the rule is pullets for winter layers and hens for breeders. The reason for this is that pullets in most breeds give more eggs than hens, and also usually do not want to sit as frequently, while the hen lays a larger egg and the chicks from them are larger and sturdier than from pullets. In some breeds the two-year-old hen lays quite as well as pullets, so I would advise you to save two-year-old hens for mothers, for your flock next year, especially if they are pure bred, and to mate them to one or more, according to the number, vigorous, pure-bred cockerels. You had better sell off all the other cockerels, or keep them by themselves and eat them, or you might have them caponized, if you can find anyone to do it for you. The usual price for caponizing is from five to ten cents per head.

Teaching Them to Roost

It is sometimes difficult to persuade the young chickens at this time of the year (September), when moved to winter quarters, to go into the coop or house, which they should occupy. The little perversities insist on returning to the place where their mother nas raised them, or they will huddle together on the ground, while the older ones fly into the low trees. Night after nigt, they have to be carried to their house. I, however, have found that by driving them gently with a broom for two or at most three nights, they will soon learn what is expected of them. A broom is by far the best way of driving chickens without frightening them.

A broom in each hand is the best way of driving a large herd of turkeys, also, by gently waving them on each side. They will be afraid of the broom, but never become wild or afraid of the attendant in this way. It is entirely possible to drive the profits out of a flock of hens by stoning and pelting them every time they get into mischief. Be quiet in your manner if you wish to be successful with hens. Make the fowls feel that, when you are present there is a protector among them, not something that is likely to scare or harm them. The only way to keep your fowls on good terms with you is by keeping them tame and treating them in a common-sense manner.

The Dry Hopper

In the matter of feeding hens on a farm, I would much prefer the dry hopper method, keeping one hopper full of mixed grains and one hopper with beef-scraps or granulated milk, and letting the fowls have free range until it is time to put them in their winter quarters. Then instead of only grain in the hopper, make the mixture of bran, corn meal and alfalfa meal, or take one of the good balanced rations sold at the poultry supply houses for the hopper. The reason for this change which should be made gradually, is that the fowls being confined, do not get the exercise and consequently may get over-fat from eating the whole grains, while the finely ground food has to be eaten more slowly. For fowls in confinement besides the hopper or finely ground feed, they should have a scratch pen in which the grain is thrown every morning for them to scratch in. This will give them the exercise which they would otherwise miss after being on free range all the summer.

After getting the fowls accustomed to their winter quarters, you can, if you wish, let them out for two hours before sun down to run on the grass or green winter wheat, or alfalfa. This will give them a little exercise and change, but it is not absolutely necessary unless quite convenient. Of course they must be supplied with green food and a balanced egg ration.

By studying the scientific and practical management of poultry and remembering the three conditions of egg production, comfort, exercise and the proper rations, you cannot fail to make a success of poultry raising on your farm.

If you decide upon making eggs for the table or market your principal object, I would strongly recommend you to have an egg route in your nearest city, taking the eggs in yourself to special customers. Your surplus fowls you could also dispose of to private customers, or if you did not wish to have the trouble of dressing them, you could send them to one of the markets. There are so many different ways of making money, if you only know how. Study that way and give your customers of the best. You will surely make a success of it.

A Practical and Inexpensive Trap Nest

The Trap Nest

"We are extremely new to the business of scientific poultry raising and have a very hazy idea of some of it. We want to develop a flock of heavy layers and would like to know what 'trap-nesting' means and how it is done." These words from one of my correspondents suggested a talk on the "trap-nest."

Trap-nests are one of the inventions of this progressive age. It is the surest, quickest method of securing better eggs and more of them. A trap-nest is a nest box, the entrance to which closes automatically when the hen steps into the nest and keeps her in the box until the person in charge releases her, thus showing which hen laid the egg.

The progressive farmer or dairyman knows that he must test the milk of his cows and he finds when he begins to do so that he has cows in his herd that do not pay for their keep. It is the same in the poultry business; in every flock of hens there are idlers that do not pay for their feed—they lay so few eggs that their owners are out of pocket by keeping them. I would not have believed this had I not discovered it to be the case with some of my own hens. The first season that I used trap-nests I found a hen which went on the nest every day but only laid four eggs in one month, while another in the same yard laid twenty-nine. It was a revelation to me. The first year I discovered that nearly one-fourth of my hens barely paid for their board. That was not the kind of hens I wanted. I was in the business for profit and not loss, so I weeded them out, and very good eating they made.

The second year I got, with a reduced flock, a twenty per cent less feed bill and fully twenty-five per cent increase of eggs—more eggs at less cost. Surely the trap-nests repaid me for the slight extra trouble of attending to them. They were not only of use in discovering the best layers, but I became better acquainted personally with each hen. I found that the hen which laid the most eggs had the most fertile eggs, while the poor layers' eggs were not nearly so fertile.

Trap-nests make the hens tame and tame hens lay more eggs than wild hens. Some hens may at first object to being handled, but after a few days they become reconciled to it. My White Plymouth Rocks were so tame that when I opened the door they would step into my hands or sit quietly until I lifted them up to ascertain the numbers of their leg-bands.

In order to make the use of the trap-nests efficient we must be able to know each hen individually, and for this purpose each hen must wear a leg-band, a small bracelet, made of copper or aluminum with a number on it.

By means of the trap-nest one can discover any hen that is becoming too fat, or too thin and she can be moved into another and more suitable pen. The trap-nest also renders a great service in detecting the egg eater. If there is reason to suspect a certain hen of this villainous habit give her an egg while she is on the nest; if the egg after a time disappears it is pretty good evidence that the culprit has been discovered and decapitation should be the verdict.

Another advantage in using trap-nests is that it gives one an opportunity to examine the hens for vermin, and by taking a small can of insect powder around occasionally while visiting the nests, and powdering the hens, they can be kept perfectly clean with very little trouble. I use a baking powder can, having perforated the lid, making a large pepper pot. A liberal use not blown on out of an air gun, but freely peppered on the hens, is very beneficial.

A Group of Four Trap Nests in Position.

I visit the nests about three times during the morning to release the hens and gather the eggs. One trap-nest is required for every three hens. When a hen is taken from her nest the egg is marked with her leg-band number and the date and credit is given her on the record sheet or record books. This is a sheet or page marked off in squares of thirty-one days with the hen's name or number at the head of the line. I mark B for brooding, S for sold, M for marketed, and so on, and have in this way the history of each hen at a glance.

Trap-nests have taught me which hens lay the best shaped eggs, which the largest size, which the strongest fertilized, which are the best winter layers, which pullets begin early, the number of eggs they lay in succession, the number of times they become broody and many other facts that can be learned in no other way; in fact, I find my records exceedingly interesting and profitable reading. Trap-nests were a perfect revelation to me and aided me in my success with poultry.

There are a number of trap-nest plans, also trap-nests, on the market, ranging in price from $1 to $25. I have bought and tried several and find that the most satisfactory trap-nest is one that has two compartments and opens in the front to take the hen off. In other words, it must be comfortable for the hen and for the attendant.

The nest box here described was made by G. M. Gowell, agriculturist of the Maine experiment station, after a careful study of the various nest boxes on the market, and is intended to combine their excellences and avoid their defects.

This is the box that is illustrated here and the description is in Mr. Gowell's own words: The nest box is very simple, inexpensive, easy to attend and certain in its action. It is a box-like structure and is twenty-eight inches long, thirteen inches wide, and thirteen inches deep—inside measurements. A division board with a circular opening seven and one-half inches in diameter is placed across the box twelve inches from the back end. The back end is the nest proper. Instead of a close door at the entrance, a light frame of inch by inch-and-a-half stuff is covered with wire netting of one-half inch mesh. The door is ten and one-half inches wide and ten-inches high, and does not fill the entire entrance, a space of two and one-half inches being left at the bottom and one and one-half inches at the top, with a good margin at the sides to avoid friction. If it filled the entire space it would be clumsy in its action. It is hinged at the top and opens up into the box. The hinges are placed on the front of the door rather than at the back or center, the better to secure complete closing action.

The "trip" consists of one piece of wire about three-sixteenths of an inch in diameter and eighteen and one-half inches long, bent as shown in drawing. A piece of board six inches wide and just long enough to reach across the box inside is nailed flatwise in front of the partition and one inch below the top of the box, a space of one-fourth of an inch being left between the edge of the board and the partition. The purpose of this board is only to support the trip wire in place. The six-inch section of the trip wire is placed across the board and the wire slipped through the quarter-inch slot and passed down, close to and in front of the center of the seven and one-half-inch circular opening. Small wire staples are driven nearly down over the six-inch section of the wire into the board so as to hold it in place and yet let it roll sidewise easily. When the door is set, the half-inch section of the wire marked "A" comes under a hardwood peg, or a tack with a large round head, which is driven into the lower edge of the door frame. The hen passes in through the circular opening and in doing so presses the wire to one side and the trip slips from its connection with the door. The door promptly swings down and fastens itself in place by its lower edge striking the light end of a wooden latch or lever, pressing it down and slipping over it. The latch is five inches long, one inch wide and half an inch thick. The latch acts quickly enough to catch the door before it rebounds. The double box with nest in the rear end is necessary as when a bird has laid and desires to leave the nest, she steps to the front and remains there until released.

With one section only she would be very likely to crush her egg by standing upon it.

The boxes, which have no tops, are arranged in cases in groups of four and slide in and out like drawers. They may, of course, be used singly by simply providing a cover for each box. When a hen has layed, the nest is pulled part way out or the cover lifted, as the case may be, and the hen removed.

I have made nest boxes myself from these plans. I used wooden shoe boxes or cracker boxes and easily made two in a morning. The wire was a little difficult to bend, but a boy did it for me. One word of caution: It is well to have nests enough, because the hens must be coaxed to lay and when they get ready they must not be kept waiting. If a hen is dissatisfied with her nest she may hold her egg for twenty-four hours and in time be taught to lay only every other day. It is wise to encourage the hens to lay and I have found these trap nests so cleverly invented by Mr. Gowell are much liked by the hens, while others I bought frightened the hens and prevented their laying. They were enclosed on the nest, pushing their heads out and trampling on the eggs, breaking some and entirely defeating the object of the nest which is "more eggs and better hens."

Grit and Gizzard

One of the most important things necessary for the health of poultry is a supply of grit of the right kind. Nature provides a use for every organ of the body, and in every body an organ for each specific duty. Most animals are provided with teeth to enable them to prepare their food for the action of the fluids secreted by the stomach, pancreas and liver. It will also be remembered that besides being crushed in the mouth by the teeth, the food is acted on by the saliva.

Nature has not endowed birds with teeth, but it has provided a good substitute in the gizzard. This is a tough, strong, muscular organ, so situated in the body that all food taken into the mouth must pass through it. Previous to passing through the gizzard, all food has been received into a pouch or bag, the crop, where it remains some time. There it is soaked with and acted upon by a fluid secreted in and by this pouch, and a modified process takes place similar to that of the saliva in the mouth of animals with teeth.

The food gradually leaves this pouch (the crop), passes through the proventriculus and into the gizzard, where it is ground up, and thence it goes to the intestines, where, after being mixed with other fluids, it passes on and the nutriment is absorbed. No doubt a bird may be made to exist for a time, perhaps a considerable time, without grit, just as a person may live for years with bad teeth, or perhaps with none at all. We all know how little such people enjoy their food or health, and surely if the birds do not have the means of masticating their food they can neither be healthy nor enjoy their food, and will not give their owners a good return for their food and care.

The Best Grit

The gizzard is a marvelously strong little mill and when provided with the proper grit, or little grindstones, will keep the fowls in good condition. Hard, sharp substances are necessary, such as flint stones or granite pounded up. Broken china, earthenware, glass and all such substances broken up make excellent grit.

When the grit has not sharp edges, the harder parts of the food are not digested, husks and green good accumulate and frequently cause a stoppage between the crop and the gizzard, so that nothing but liquid can pass. A lack of sharp grit brings on diarrohea; also, the gall overflows and sometimes the gall-sack bursts. There are two passages, one into and the other out of the gizzard; they are both on one side of it. The one leading out of it is much smaller than the one leading into it. Thus the gizzard can receive larger substances but cannot get rid of them until they are ground small; and sharp grit is needed for this.

When I first came to California I purchased a gristmill and, alas, I had broken china also! I had two dozen hens just bought and proceeded to grind up some crockery for them. The man who was building my fence thought it dreadfully cruel of me, remarking, "It's euongh to kill a dog; let alone those poor hens." "The hens will not eat it unless they need it," was my reply, though I agreed with him about the dog. To his surprise those hens ate almost a quart of it. None of them died and they soon commenced to lay. Give the little chicks the small chick-grit. Eight pounds of this will be sufficient for the first two months of the life of fifty little chicks and then they should have a larger size. One hundred pounds of hen grit, which can be bought at the poultry supply houses, is sufficient to last a hundred hens about a year.

Pigeons consume more grit than hens, proportionately to size. Give pigeons grit to keep them healthy. My attention to grit and gizzards was aroused many years ago. "Will madame look to what I have found in the interior of this fowl?" said my French maid to me. She had opened the gizzard of a fat young hen and had found thirteen china buttons and two pearl buttons or parts of them, mixed with the black adobe mud. Since that day I have tried to keep my fowls well supplied with grit.

Starve for Lack of Grit

"I cannot think what ails my fowls," said one lady. "They have all the food they can eat, but here is another dead." "Have you ever opened one to discover the trouble?" I asked. "Yes, but I never find anything." "Well, I think your fowls have indigestion," I said, "but we will hold a post mortem on this one and try to solve the difficulty." We fund a medium sized gizzard, full of dark earth, no stones, no grit, not even buttons. That told the story, the fowls were starving to death in the midst of plenty just for lack of grit to grind their food.

I occasionally make curious discoveries when I hold a post mortem, for the contents of a school boys' pockets are scarcely more varied than those of a fowl's gizzard, when not supplied with the proper kind of grit. My Indian Runner ducks being great pets and never doing any mischief were allowed the freedom of my place. I had noticed them around the out-door fireplace where the cauldron was boiled; old boxes, building scrap and rubbish being used for the fire.

I thought the ducks were picking up bits of charcoal, but one morning I found a fine duck dead. The post mortem revealed an enormous gizzard, twice the usual size, on opening which I found a number of nails, some bits of wire, two two-pointed tacks. Several of the nails were embedded in the gizzard and the largest one pierced quite through it. The ducks had always been supplied with plenty of river sand, but this particular duck seemed to have developed an ostrich's appetite. After that I gave them also the smaller chick grit and with most excellent results, for never ducks laid as many eggs as did those. Grit, oyster shells, or clam shells, and charcoal are indispensable for fowls.

The Symptoms of Grit Craving

When your hens seem "mopey" just break up some old china, and see if they will not refuse the best food for it.

When you see water run from a hen's mouth, when she puts her head down, the trouble is indigestion. Give her grit and charcoal.

When your hens do not care for their food, tone up their appetites by a dose of grit.

When they are not laying as well as you think they should, give them grit.

When hens moult slowly it is often from impaired digestion. Give them grit and charcoal.

When you want the hens to derive all the benefit of the nutrition in the food, supply them with good, sharp grit.

If you want vigorous, profitable hens give them a liberal supply of grit.

When your hens are too fat, when they lay thin shelled eggs, give them grit.

A friend of mine was very much troubled with soft-shelled eggs. She got her husband to take his wagon to the hills, where there is a good quarry of what is called rotten granite. He brought home a load of it and in a few days the hens laid hard-shelled eggs and she told me that the shells were so hard that the chicks could hardly break out of them.

The value of good sharp grit can scarcely be overestimated, and yet even intelligent people do not realize it. Some think that there is grit enough in the natural soil. This is rarely the case, for hens, wild birds, or pigeons pick up the sharpest and best grit, so that even on a farm where the hens have free range there is rarely enough grit of the proper kind, and when fowls are kept yarded there is never enough unless they are artificially supplied. If you doubt this, try the experiment of giving your hens some broken china. The pieces should be not larger than a pea and should have three sharp corners. You will be surprised to see how eagerly the hens will eat the china.

The best layer I ever had laid 225 'eggs in nine months and moulted during that time. She was the greatest eater of grit I ever saw. Every night before going to roost she ran down to the grit box and took three pieces. Every time she laid an egg she refreshed herself with some grit, and I learned by observation that all my best layers were the most constant visitors to the grit box. Hens that consume the most grit are those that get the most nutrition out of their food, lay the most eggs, are the healthiest, have the most fertile eggs and pay the best.

Grit to grind the food and charcoal to keep it pure during this process and, for laying hens, oyster shells to supply the lime for the eggshells, these are so necessary that we are almost tired of the mention of them in the poultry papers, but "lest we forget" I have written about them again.

Pests of a Poultry Yard

Fleas

The common hen flea (pulex avium) is prevalent in the Pacific States. It is found in filthy hen houses, especially those located on sandy soil. Dirty nests, cracks, dust and dark corners are favorite breeding places for them. They produce great irritation of the skin and in young birds the growth may be permanently stunted and many young chickens killed by them.

For treating flea bites bathe the bites with vinegar and water, or lemon juice, and apply carbolated vaseline or lard in which a little carbolic acid has been mixed—5 drops of carbolic acid (90 per cent) to a tablespoonful of lard.

To free poultry houses and yards of the fleas, use whitewash freely, adding a pint of carbolic acid to every twelve gallons of whitewash. Spray it or slop it thoroughly into all the corners and cracks. Dark dusting places in the poultry yard afford favorable breeding places for fleas. These corners should be soaked with hot soapsuds or boiling salt water to kill the young broods of fleas. Use carbolized lime, tobacco dust, and moth balls in the nests.

Bedbugs and Ticks

Bedbugs sometimes attack poultry on their roosts and suck their blood. In California there is also a species of tick that is fatal to poultry which somewhat resembles the bedbug of the East. To destroy them fumigation is usually employed, either fumigating with sulphur or, better still, the cyanide process used for the scale on citrus trees.

To fumigate with sulphur close every door and window and see that there are no cracks to admit the air. Burn one pound of sulphur for every 100 square feet of floor space in the house. A house 10x10 will require one pound of sulphur: one 20x10, two pounds, and so on. The sulphur must be burned in iron vessels which should be set on gravel or sand so there may be no danger from fire. Into each vessel put a handful of carpenter shavings saturated with kerosene and upon these sprinkle the sulphur. Apply a match to the shavings and hastily leave the house, closing the door. The house should remain closed for 5 hours. Fumigation may be followed by thoroughly whitewashing the inside of the house. Painting or spraying the house with corrosive sublimate is also very effective. Care must be used in handling this poison.

Mites

There are several varieties of the tiny blood-sucking mites to be found in carelessly kept henneries. The red mite is the most common and active of all parasites which attack birds. It is about one thirty-fifth of an inch in length, white or grey in color except when filled with blood, when they will be red or black. It hides by day in the corners and crevices of buildings, nests, perches, floors, etc., where they may be found in clusters. At night these clusters scatter over the birds and by pricking the skin, can fill themselves with blood. They are injurious not only on account of the blood they draw, but because of the itching pain and loss of rest. They will even kill young fowls and setting hens. When they are discovered, vigorous means should be adopted to get rid of them. The Iowa State Experiment Station gives a full description of the best and cheapest way of exterminating these mites. At this station the kerosene emulsion was found to be perfectly effective in killing them. It is made as follows:

KEROSENE EMULSION—In one gallon of boiling water dissolve one pound bar of soap or one pound of soap powder. Remove from the fire, add immediately one gallon of kerosene, churn or agitate violently for ten minutes, or until the solution becomes like a thick cream. If the oil and water separate on standing, then the soap was not caustic enough. Take one quart of this, add to it ten quarts of water, spray thoroughly the houses every three days with this diluted emulsion until all the mites are exterminated. To make it more effective you may add one pint of crude carbolic acid to the emulsion as soon as taken from the fire. The diluted emulsion (one part to ten of water) is also used to rid fowls of lice. By using this spray once a month always, the houses can be kept perfectly free from vermin and thoroughly disinfected from disease.

Lice .

There are nine varieties of lice affecting poultry. Some of these lice spread rapidly. One infested bird is capable of spreading the vermin through a large flock. They cause dumpishness, drooping wings, indifference to food and may stunt or even kill the chicks. One of the best means of preventing lice is the dust bath. This bath should be a wallow of freshly turned earth, mellow and slightly damp, out of doors under some tree in the summer time or in a box six or eight inches deep in the hennery in the rainy weather. Provided with a good dust bath, healthy hens will almost keep themselves clean from lice. When fowls are badly infested with lice they should be well dusted with a good lice powder, of which there are a number on the market. Two good powders can be made as follows: To one peck of sifted coal ashes add one-half ounce of 90 per cent carbolic acid. When mixed thoroughly, add an equal amount of tobacco dust. 2nd: Take half peck of sifted road-dust, four fluid ounces of any good liquid lice killer; mix thoroughly and add bulk for bulk of tobacco dust.

The roosts may be painted with liquid lice killer, or the fowls placed in a box for three hours, the floor of which has been painted with lice killer and the top covered with burlap, care being taken not to smother the hen. The nits of lice hatch about every five days. The treatment should be repeated until all the young lice have been exterminated.

Diseases of Poultry

There is no reason for chickens being unhealthy except, as a general thing, from the carelessness or ignorance of their owners. Carelessness in not keeping the fowls clean, in not being regular in their feeding, in the lack of pure water and shade and in giving them either draughty sleeping quarters or too close and badly ventilated coops.

Poultry keepers in the East, after years of trouble and anxiety over roup, which I really think is much worse there than here, are coming to the conclusion that open front houses even there where they have zero weather, will prevent roup and colds.

Here in our favored climate, open front houses, cleanliness and plenty of green food are a sure prevention of roup.

I am glad to be able to say that although there are more than double the number of pure bred fowls in California now than ever before, there is a minimum amount of roup. Poultry raisers are using common sense in the feeding and care of chickens, looking upon poultry raising as a business, a money proposition, when handled in a business-like way, and the result is very little roup and less sickness of any kind.

Roup must be transmitted by contagion; healthy fowls will not have it unless a roupy fowl is introduced into the flock, or the infection is brought in through water or food, through coops in which roupy fowls have been confined or through the infection being carried on the garments of the attendant.

Many Kinds of Roup

It was formerly the custom to call nearly all the ailments of fowls due to taking cold by the name of "Roup." Dr. Salmon of the Bureau of Animal Industry, Washington, D. C., makes a distinction, however, between the different kinds of colds or roup, simple catarrh, and infectious catarrh also called roupy catarrh, and diphtheric catarrh or diphtheric roup. Simple catarrh is easily cured, will often get well without treatment; roupy catarrh is very infectious and more difficult to cure; but diphtheric roup is the worst of all and greatly resembles the diphtheria of children. There is also another disease called "Canker" which much resembles diphtheric roup, but is less severe. It is caused by another germ and needs other treatment.

Catarrh

All of these diseases commence in the same manner. Usually the first symptoms noticed are a slight discharge from the nostrils, eyes wet and watery from mucous, and often some bubbling at the corners with coughing and sneezing. In simple catarrh more serious symptoms will not have developed in a few days, but with roupy catarrh the discharge thickens and obstructs the breathing by filling the nostrils and there is a foul odor to it. Sometimes swell-head develops, then one or both eyes are closed, the birds wipe their eyes on their shoulders, sleep with their heads under their wings and the discharge sticks to and dries on their feathers. This dried mucus will spread the disease through the flock, for in it are the germs of the disease, the seeds of which may be sown whenever the chicken moves or shakes itself, or when others touch it or a feather falls. Chickens with this disease should be isolated, the mucus gently washed off, using a disinfectant in the water, a few drops of carbolic acid or a tablet of protiodide of mercury in a pint of water. Roupy catarrh is difficult of cure, is very infectious and often fatal.

Diphtheric Roup

Diphtheric roup is the worst of all. It requires different remedies to the simple catarrh or roupy catarrh. It commences usually in the same manner with a slight cold, but the mucus membrane of the mouth, throat, nasal passages, and the eyes are affected. False membrane forms on these parts, very much resembling in appearance the diphtheria of children, and by some, thought to be the same. At first the patches are small and scattered but have a tendency to run together. The disease appears suddenly, the fowl is feverish, dumpish, and disinclined to eat. As the disease progresses the mouth and throat become filled with false memberane and mucus until the fowl dies of suffocation, or the poison from the disease gets into the circulation and the fowl dies of blood poisoning or paralysis.

Canker

Canker is sometimes confounded with diphtheria. It is an ulcerative disease of the mouth. It is frequently found in cock birds after fighting and is common in birds that have been working in mouldy or musty litter or that have been fed on spoiled grain. The disease is seldom noticed until the fowl shows a collection of yellowish ulcers or cheesy growth on the roof of the mouth, the side of the tongue or the angles of the jaws, and sometimes at the opening of the windpipe. It is very common among pigeons.

Roup cures can be bought at the principal poultry supply houses but for the use of those living in the country too far away to procure these, I will give a few simple remedies that can be easily and quickly used in the first stages, thus arresting an apidemic. For local treatment a good atomizer is the most satisfactory way of applying it, or a small syringe, and as handy as anything is a small sewing machine oil can.

Remedies

(1). When first the cold is noticed, put a bit of Bluestone (sulphate of copper) in the drinking water. A piece as big as a navy bean in a quart of water, not any stronger. This is a germ killer, dries up the cold in the head, is a disinfectant and will prevent the other chickens taking the disease. So, if any chick takes cold put this into the water of the whole flock for a week to prevent the dsease spreading.

(2). For a Common Cold: A pill of quinine and one of asafoetida

(1 gr. of each) with half a teaspoonful of cayenne pepper will frequently cure a cold in one night. Aconite also is a good remedy. One drop in a teaspoonful of milk. Always give a grown hen the same dose as to adult human beings.

The following are cures for Roupy Catarrh:

(3). One tablespoonful of castor oil, half a tablespoonful turpentine, a tablespoonful of kerosene, a tablespoonful of camphorated oil and four drops of carbolic acid. Shake before using. Squirt a drop up each nostril and into the cleft of the mouth, and for swell head rub the whole head with it. This is an excellent cure and cheap.

(4). Put one cupful of kerosene in half a gallon of water; the oil will float on top; dip the fowl's head slowly into this, holding it under whilst you count three. It will sneeze and cough and you must wipe off all the mucus with a rag and carefully burn the rag. Repeat the treatment twice a day.

(5). Take of lard two tablespoonsful; vinegar, mustard, cayenne pepper, each one tablespoonful; mix thoroughly, add flour enough to make a stiff dough. Give a bolus of this the size of the first joint of the little finger. One dose frequently cures. If not, repeat in twelve or twenty-four hours.

(6). Dr. N. W. Sanborn gives as a remedy: "Spray all mucus surfaces with the following: Extract of Witch Hazel, four tablespoonsful; liquid carbolic acid, four drops; water, two tablespoonsful. Do this twice a day, squeezing the bulb of the atomizer five times for each nostril and twice for the mouth If there is any watery or foamy eyes, give one squeeze for each.

(7). One part of pulverized gum camphor and seven parts of pulverized liquorice root. Blow up the nostrils, into the cleft of the mouth and down the throat. This should be made fresh as the camphor evaporates.

(8). Equal parts of powdered alum, magnesia and sulphur blown into the throat and nostrils through a quill.

(9). For Diphtheric Roup: Peroxide of hydrogen is, I think, the best remedy. Dilute with from one to three parts of water. The solutions, when applied to diseased surfaces, begin to foam, and should be repeated until there is no more bubbling. A little of the solution forced into the nostrils by the use of a dropping tube or atomizer is driven higher up into the nostrils by the force of the foaming, reaching parts otherwise out of touch.

(10). For Canker: Four grains of Sulpho-carbolate of zinc to one ounce of water. Paint the canker spots with this night and morning and in three days the germs will be destroyed. The chickens should have nourishing food, such as bread and milk and chopped onions.

If you have any doubts as to whether the disease is canker or roup you had better use the peroxide of hydrogen one day and the zinc the day following, alternating the treatment. It will not do to mix the two medicines at the same time as one neutralizes the other.

Town-Lot Fowls

The rear of a city lot can be made to yield both profit and pleasure when devoted to poultry and fruit trees, and many families may enjoy fresh eggs and an occasional roast chicken, or a "Christmassy" chicken pie by simply utilizing some of the vacant space in the rear yards of their homes.

We sometimes hear that chickens cannot be raised successfully on a city lot because the land is too valuable and that the business will not pay where all the food has to be bought.

The value of a city lot is often over-estimated when chicken raising is suggested for the back yard, but the question is, what income is your back lot now yielding?

I expect that the majority of city back lots are either an outlay or an eyesore to their owners. They grow nothing but grass or weeds, for which nothing is received. When mowed there is that expense to it, with the water tax added, which is not inconsiderable.

As much as I like lawn and flowers in the front of the house, I think the oftimes neglected back yard should be made valuable also. Nothing to my taste can improve it like fruit trees, which are benefited by having poultry around them, and will bring in good returns, as I know by experience.

The main requisite to making a success of poultry raising on a city lot, or anywhere else in fact, is to be thoroughly in love with your fowls and your trees. The man or woman who hates to work around the hens, who grudges the time and trouble, will never make a success of the work and had better let it alone.

How to plan your back lot? It should be fenced to suit your

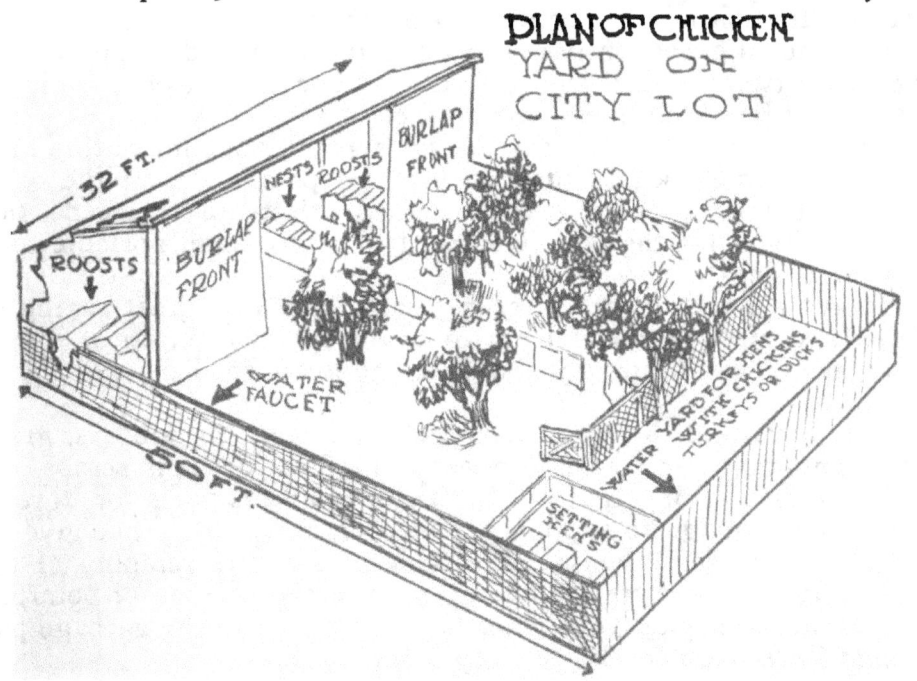

PLAN OF CHICKEN YARD ON CITY LOT

space and poultry. If it is a small yard, it may be difficult to fence it high enough for the active breeds such as the Leghorns, but if you use poultry netting and do not place any rail on the top, you will not have any trouble with the American breeds, even with a comparatively low fence. If there is no rail on the top, the fowls do not see where the netting ends and they seldom try to find the top, but with a rail they light on that and over they go.

It may help a beginner to see the plan of my chicken yard on a city lot. The chicken yards are 50 feet by 32 feet; there are eight fruit trees and three water faucets in the yard. The fruit trees, plum, peach and fig, yielded several dollars' worth of fruit two years after planting, and as they grew older, increased the value of the crop in the back lot, and gave the fowls shade.

Hen House Construction

The earth around the trees is kept well spaded and moist, so the hens enjoy it as a dust bath and that keeps them clean from lice and mites. The hen house is a shed thirty-two feet long and eight feet wide. It is divided in two parts for two pens of fowls. Each end of it is composed of a roosting room eight feet by eight feet, with space enugh for forty hens, if necessary, although I never wish to keep more than twenty-five in each side.

The roosting room is separated from the scratching pen only by a board twelve inches wide, to keep out the straw. The back and sides of the roosting room are of tongued and groved flooring and perfectly tight. The whole length of the front of the shed is open, except the roosting room, which has a front of burlap. One side of the roosting room is entirely open into the scratching pen, so that the roosting room is only tightly enclosed on two sides and has free ventilation into the scratching pen and only the burlap on the south side. Consequently my fowls never have colds. The roof is of shakes twelve inches to the weather. The back of the shed is six and a half feet high, the front five feet.

At the south end of the two yards is a smaller one for setting hens or for young chicks, as they do better kept away from the older fowls. This small yard is very useful for fattening chickens, turkeys or ducks for the table, and in it I have a small portable coop for the youngsters.

I have a water faucet in each yard. This is a great saving of labor and anxiety, for if I am to be absent any length of time I leave the faucet dripping just a little and know the hens will not go thirsty.

I feed grain in the scratching pens, dry mash in hoppers, green lawn clippings and refuse vegetables, besides the table scraps.

There is a saying that an American family wastes or throws away food enough to support a French family. Why not give all this waste to some hens? The table scraps, the scraping of the plates, the outer leaves of cabbages, even the parings of potatoes, apples and nearly all vegetables now consigned to the garbage pail would be enough to almost keep a few hens.

Possibilities of a Town Lot

Have you any idea what returns one dozen laying pullets or hens would give you? I have, for I have kept that number on a town lot. I have not an accurate account of all the eggs laid, but I know there were over two thousand in one year, more than enough to supply a family of six with delicious fresh eggs and to raise between fifty and sixty young fowls for frying and roasting, besides the old ones for stews or for "poulet au ris," a French dish of which we are extremely fond.

Nine-tenths of the home owners have sufficient space in their back yards to produce enough chickens and eggs to supply their own families, and in this way greatly lessen the expense of living, or in other words, make enough to pay their meat and grocery bills, or else give them all the fresh eggs they can consume with a nice fry always available for Sunday dinner or when a friend unexpectedly drops in.

I will give you a formula for feeding hens on a town lot which I will guarantee will give you eggs in abundance and at all seasons. It is easy to feed, for all you have to do is to mix it dry in a big box and dip up half a bucket, once or twice a week and fill a box or hopper full of it as the need is. It is quite dry and will keep any length of time.

Formula for Balanced Ration

Mix by measure two parts bran, one part corn-meal, one part alfalfa meal, one part beef scraps. Keep some of this in a box or hopper or bucket—dry, perfectly dry—always before the hens. This dry food in the hopper lasts quite a long time, for the hens prefer the table scraps which are fed to them only once a day (at night) and they like lawn clippings, but this dry feed keeps them in just the right condition for egg production—neither too fat nor too thin.

If you do not want to take the trouble to mix this for yourself, you can go to any of the poultry supply houses and buy the food already mixed. This food when put up by reliable firms is what is called the "balanced ration"—that is, it contains the elements of the egg—and when the hens are fed this they simply cannot help laying. They are egg machines which turn the properly balanced ration into eggs.

Town Lot Youngsters

The Moulting Season

The moult with hens in the natural state lasts from sixty to a hundred days, but with some hens, especially with hens that have hard, close-growing feathers the moult and the results of it will sometimes last over a hundred and fifty days; in fact I have known of some that went six months without laying any eggs. Too long to spend half a year dressmaking. Think of the loss to their owners! I did not wonder at the man who told me of it, saying that he just turned them out and "let the blamed things rustle for themselves," but I thought if he had helped them "rustle" perhaps they would not have been so long about it.

Let us consult Nature as you know I am very fond of doing. After the wild bird has raised her young and her responsibilities are somewhat over, she moults. The older she is the longer and slower is the process of dropping her feathers and growing them again, because as she ages her vitality is gradually lessening. It is the same with hens; the older a hen becomes the longer will be the period of the moult, and not only that but the later will it commence. Let us again turn to Nature and in this copy her. We want the old hens, if we keep them at all, to be the parents of our young next spring and we are only keeping them over for a certain reason (or for sentiment), as they have, perhaps, proved themselves to be our very best layers, or as the parents of our prize winners, or may be prize winners themselves and therefore we want their offspring in the hopes of perpetuating these excellent traits.

The Starving Process

How shall we help these elderly hens to get quickly through the moult? Some years ago I read of a man in New York State, who claimed he could make his hens moult at any time of the year and therefore he could also, by controlling the moult, make his hens lay at any time of the year. His plan was to starve the hens and so stop their laying and when they had stopped for a week or two he fed them highly with fattening food. This he said made them moult and drop their feathers very quickly so that in a few days the hens would be almost nude and the new feathers would come in very rapidly. His theory was that when hens sit for three weeks on eggs and raise a brood of chickens they moult quickly because they grow thin during incubation, and when they have the rich feed which is given to the little chicks it makes them shed their feathers and assists the moult.

His theory sounded very plausible and I decided if he could do it I could also and tried. I discovered the New Yorker was only partly right in his deductions and that it does not pay to force Nature out of season.

The following year I was much more successful for I only attempted to "assist" Nature and not to "force" her. I did not try to make the hens moult in June, but waited till nearer to the natural time of the moult, that is, until August. I then put the hens on green food. I know that is hard to get at that time but I had lawn clippings, vegetables and melons, or even alfalfa hay cut in the clover cutter and soaked for some hours in water, and I dispensed with all the grain and meat. I kept them on this green food for about three weeks until their avoirdupois was considerably lower and most of them had stopped laying for a week.

Dipping Fowls

Meanwhile during their fast I saw that they were entirely clean from lice, either by keeping them well dusted with insect powder or by giving them a good warm bath in warm soap suds, rinsing them in a two per cent solution of carbolic acid or water and creolin or the kerosene emulsion. I have tried all of these with good success.

This washing seems to loosen the feathers and will clean the fowls of lice. If lice are left on the fowls at moulting time they eat little holes in the tender sprouting feathers and these little holes in the web of the feather will certainly bring a "cut" from the judge in the show-room, and for the whole year will tell the tale of careless handling by the owners. In washing or dipping fowls for lice there are two things to be remembered: First, do it on a bright, warm, sunny morning, so the fowls will have time to get thoroughly dry before sundown, and, secondly, see that every feather is thoroughly soaked. If you skip a feather a louse will take refuge on it and commence to breed again as soon as the hen is dry. If there are any lice the disinfectant in the bath will kill them and the warm suds also loosen the nits of the head lice. Those lice lay two silvery, white nits at the shaft of the feather and it is difficult to get them off.

Mature hens which are fed sparingly for about two weeks and then receive a rich nitrogenous ration, moult more rapidly and with more uniformity and enter the cold weather of winter in better condition than the fowls fed continuously during the moulting period on an egg producing ration.

What to Feed

It is largely a question of what not to feed as well as how little to give the birds you wish to moult early. There is one line of foods that you may feed in unlimited quantities, and that is the green vegetable, the waste, small beets and thinnings of the garden rows can be supplied every day. My own plan in the days when I had small ungrassed yards, was to give full quantities of lawn clippings, putting them into the yards an hour before dark. This gave the birds time to fill up at night and yet the uneaten clippings would be still fresh in the early morning. If you have had no experience in the use of lawn grass you will be surprised to see how much a few hens will eat. If your hens have very large yards with fruit trees to supply some falling apples or pears, the birds will do very well without other food. We are inclined to over feed our birds with grain in the warm weather and, unless the food is really much less than usual, you will fail in getting an early moult.

This low feeding or starving process as it is called by many, is the important factor in the forced moult. Unless you really do this in good shape the birds will continue to lay and will shed their feathers in mid-autumn.

Handle your birds on the roost to test their weight. They must be thin in body, yet good in color of comb and wattles. I find that birds take from fourteen to twenty-one days to get real thin. You will notice as you put this plan into practice that the egg yield will drop off until no eggs are being layed; that the birds are on the run all the day long, coming to meet you at any point of the fence you may approach. The birds show that they miss some of their usual food. This thinning will do no harm to the birds; in fact it adds to the health of the birds for months to come.

The Full Ration

When the birds have lost all superfluous flesh, when the eggs have ceased to appear for a week, feed them good, full rations of growing foods. Now is when you add meat, beef scraps, green bone, cornmeal, and linseed meal. You can give them a morning meal of two parts cornmeal; three parts bran, one part beef scrap. At noon feed a small handful of wheat or barley to every bird and at night a full feed of wheat or corn. Do not neglect to furnish full supplies of green food and vegetables all the fall.

The change from the low feed to the full rations will be followed by the rapid dropping of feathers. The feathers will fall off all over the birds so that many of them will be almost naked. This result will be seen in most of the birds. A few will fail to respond, more if you do not follow the plan as outlined.

Keep the full feed up until the birds get the new coat of feathers and begin to lay a few eggs. Then feed them as you do the fully mature pullets; avoid feeding of heating foods (corn and corn products) lest you start another moult in the late autumn.

The forced moult is ONLY FOR MATURED FOWLS, or fowls that are over a year old. You must not starve the pullets. You must keep them growing. They will stand more heating food than hens. Let the pullets do most of your winter laying, but do not neglect anything that will induce the older birds to give you a good share in the profits of winter eggs.

To sum up the whole matter in a few words, if you want to hasten the moult, do not try the experiment with all your fowls, but take a few, separate them from the others and about the middle or end of August, commence to shorten their food. You can do this suddenly, giving them only green food and all the green feed they want. Secondly, keep this green feed up for two or three weeks, or at least one week after they have stopped laying. Thirdly, the green food should be clover, lawn-clippings, alfalfa hay cut in a clover cutter and soaked in water; beet tops, cabbage, lettuce, etc. Fourthly, after the three weeks' fast, feed rich food, fattening food, sunflower seeds, kaffir corn, wheat, barley, oats and meat. Fifthly, when they begin to lay on this food, which they will do in about a month when they have completed their coat, gradually change the food, taking away the corn and its products, and the linseed meal, and anything that would be very fattening.

Color of Feathers and Skin

The feeding of the fattening foods adds heat to the body, fever our grandmothers called it, and this fever seems to loosen the feathers all at once—just what we want—and they fall so quickly that the hens are almost nude. Then is the time for care in feeding if you have exhibition stock, for I am certain color can be greatly controlled by food.

Now, I know by my own experience that yellow corn will give yellow feathers (brassy feathers) to white fowls when freely fed; that cottonseed meal will have the same effect, for that is what we add to the fattening food the last week to give the yellow tint to the skin. I know that iron in the drinking water has the same effect with white fowls. With colored fowls, such as Brown Leghorns or Partridge fowls or Buffs the iron and the corn will intensify and make more brilliant and bright their colors.

The fowls that are making their new coats, the coats that have to last the hens a year, all need plenty of green food and grain. The white fowls instead of yellow corn, should have oats, hulled oats are best, but if you cannot get hulled oats, soak the oats in scalding water so the hulls will be softened. Hulled oats may appear to be more expensive than the unhulled, but there is so much waste, so much indigestible fiber in the unhulled oats, that I decided that it was more profitable to feed the hulled oats. For those who are feeding cockerels which they want to exhibit in the winter; for the white or black and white, give them shade, plenty of shade, for our California sun will dry out the yellow; cut off all the yellow corn and all cottonseed meal; feed oats, wheat, barley grit, charcoal and have granulated bone always before them. For the colored fowls add linseed meal to the ration. It will deepen and brighten the colors.

Town Lot Yards of Mrs. E. L. Walsh

Value of Economy

The old saying "a penny saved is a penny earned" may well apply to the poultry business. To make money in the business, one must practice economy in every direction.

Economy in Grain

First: Economy in buying the food. This is very important. The available grains vary in different places in price; in some localities, for instance, barley is cheaper than wheat, then utilize barley; that is to say if there is a decided difference in the cost, remembering that barley has a husk on it, which is indigestible fiber, and that fowls do not like it as well as wheat, although they eat it readily if rolled or soaked or sprouted, and the analysis shows the same nutritive ratio as wheat. Again in some places, oats can be obtained very cheaply, and these are a most valuable grain for feeding and building up large, sturdy frames in the young fowls, promoting egg laying and inducing fertility in the eggs. I have great faith in oats—it is good for man, beast and bird, but the husk is the difficulty there. The oats should be scalded or clipped, or better still, hulled to make them thoroughly available. In Oregon and Washington, oats are less expensive than in the south, and therefore should be freely used there. By commencing the use of them early, the chicks will be vigorous and of large frame.

Then again rice, rice hulls and rice bran are cheap in certain localities, such as in San Francisco and Seattle, where large quantities are imported and cleaned, and these can be had very cheaply and utilized either in the dry or wet mash. In other places where beans and peas are grown in quantities, the refuse of these, which is not worth marketing, can be used most advantageously.

Broom-corn seed is a most excellent food and costs very little. I had in Oklahoma many tons of this, to which the fowls had free access and with green growing winter wheat, a little milk and table scraps, they layed all through the moult and through the winter, notwithstanding the blizzards and zero weather. Nothing seemed to stop their laying, and I attributed it to the broom-corn seed. Sorghum seed is equally good.

Another little economy I found quite good among the little chickens was buying dry or stale bread from the bakeries at 25c a sack weighing 25 pounds. This I took home, cut same in slices and dried in the sun or in the oven, ground in the grist mill and used either moistened or dry, for chickens, turkeys and ducks.

Economy in Vegetables

Then, again, there are the various vegetables, many of which can be had for almost nothing. There are "small potatoes." It generally raises a smile to talk of these, but they make a most excellent addition and variety to the fowls' bill of fare. Small raw potatoes can be chopped up in the chopping bowl in a few minutes, also turnips, carrots and onions, and the outer leaves of cabbage, cauliflower or celery. I bought the largest chopping, or butter bowl, I could find, and a double bladed chopping knife, and used it every day, especially for the little chickens and turkeys. Small potatoes, turnips and carrots can be boiled, mashed, mixed with bran and blood-meal, or with milk, and make a good variety in the diet. If you have other vegetables to spare, such as beets, cucumbers, pumpkins, etc., and find the fowls do not at first like them, chop some up and mix bran with them and soon the hens will acquire a liking for them.

Another economy is using the leaves which fall from alfalfa hay. When the hay-mow begins to get empty, sweep up the leaves and put them in a box or sack to mix in either the dry or wet mash. I used to try to keep the last two bales of the alfalfa hay, as the balers would sweep up the leaves and put them in these last two and this was just what I wanted for my hens. Sometimes I soaked the leaves and fed them at noon, keeping the alfalfa tea to mix in the mash with potatoes and bran or whatever I was feeding. I always said the alfalfa tea was as good as beef tea. There are many ways of economizing in the feed.

Economy in Labor

Another thing to economize is labor. I know many a farmer's busy wife will agree with me in this. I found the dry feed a great saving of time and strength. It was much less labor to carry around to my many pens of fowls buckets full of dry food nicely mixed in the proper proportions and pour it into a box, or trough or hopper and let the hens eat it dry, instead of laboriously mixing it with water. Before trying the dry feed, I had so many hens that I had a large trough made, like a plasterer's trough, and I used to mix and turn the mash with a spade or hoe and then fill those large buckets full and put them on a child's express wagon to pull out to the pens. This was quite hard work and I hailed with joy the easier task of carrying the lighter buckets of dry food. I found, too, that it saved time to mix up the food by the sack full or bin full, then all that was required was to dip up a bucket full for each pen. I showed this plan to a friend of mine and later had a letter from her telling me it was a great comfort, for all she had to do was to send her Jap boy out to that certain box or bin and tell him to feed that; she knew he could not make a mistake for it was ready mixed.

Economy in Water

Another economy: Have a water faucet in each pen. This may seem like an expense at first, but it will pay in the end, for fresh water is as important as good food, and if it requires but a turn of the faucet the hens are sure to be amply supplied. At one ranch where there was an abundance of water, I saw a small fountain which ran into a basin and that in turn overflowed into some cobblestones and a drain, so that the hens had always fresh water without drawing on either the strength or time of their owner.

I would, however, caution chicken raisers against allowing the water to run in a stream from pen to pen, as that may carry infection, especially the infection of colds and roup. One gentleman who had 3000 fowls told me that letting the water run in a small stream through his pens, had ruined him in the chicken business. One pen at the top of the hill got roup, and the infection was carried through to all of them. In Kansas one of the worst outbreaks of chicken cholera came from a creek. All the farms on that creek lost all, or nearly all their chickens, from drinking contaminated water. A faucet in every yard would be cheaper in the end than an outbreak of roup or cholera.

Economy in Fencing

Economy in fencing came in very handily one summer. I found I could make a very good temporary chicken-wire fence with posts 50 feet apart by "darning" in a lath every eight feet or so, passing the lath in and out of the wire meshes before putting up the wire. This keeps the wire stretched and when taken down it can simply be rolled up and used over and over again, keeping the lath in it ready for the next time. I found chicken wire and lath quite an economy. I made cat and hawkproof little pens of this. Bought a bundle of six foot lath, some two- feet chicken wire and made most useful little panels six feet long with the laths, stretching the chicken wire on them and tacking it down with two-pointed tacks. I wired or tied the panels at the corners and had a larger panel go over the top made of six foot wire. I did not have to kill any cats or have fusses with the neighbors. The little panels were untied and piled up for the winter time and put in the barn, coming out almost as good as new the next season. They were cheap, light, easily handled and very satisfactory.

Beware of Spoiled Food

It is poor economy to buy spoiled grain of any kind. The best is none too good and anything that is spoiled is very apt to bring in disease. Wheat or any grain that has been moistened will develop fungoid growth; smutty wheat, etc., is almost poisonous to fowls, while, of course, we know that there is no grain that so nearly approaches the analysis of an egg as does wheat, when it is good. Corn, likewise, if it has been dampened, will commence to ferment and that will disagree with fowls. At one time there was a fire at a flour mill in Los Angeles. A great deal of the spoiled wheat was sold for chicken feed. "Anything was good enough for chickens," was the cry, and hundreds of chickens lost their lives from that wheat. The owners of the fowls thought it was chemicals that had been used in suppressing the fire, but it was nothing but water, some of the firemen told me, that had been used for extinguishing the fire. The dampened wheat became musty and mouldy and it was that which killed the chickens. Again in using beef-scraps, meat meal, blood meal or animal meal, be careful to buy the best you can get, and keep it carefully away from any dampness. Dampened or spoilt animal food is poisonous to the chickens and many a fowl has died from ptomaine poisoning from using spoiled animal food. One of the greatest economies is to buy in large quantities.

Most Suitable Green Foods

Whilst we are on the subject of economy we must not forget the two green foods that are the most suitable for fowls—clover and alfalfa.

Let those who are living on a town lot have a clover lawn; clover requires less water than blue-grass or any lawn grass in this climate, and is easily grown when once it is properly started. The lawn clippings are just the right length for green food and if necessary, the hens can be turned out on to the lawn two hours before sunset, and will then busy themselves nipping off the clover leaves; they will not have time or inclination to do damage by scratching. A run on the lawn before bedtime is a wonderful tonic for chickens that are yarded closely all the day.

Every farm should have an alfalfa patch, if not a good big field of alfalfa, and no chicken ranch is complete without one, for the youngsters should have a good alfalfa run to properly develop them.

Alfalfa is a legume; is rich in nitrogen and enriches the land upon which it is grown. It is the best green feed next to clover for the hens or cows, and the hens love it. It is equally good for ducks and turkeys. The question of economy of labor is a very serious matter in poultry raising, and by having a good alfalfa patch upon which the hens may be turned several hours daily, the labor of cutting and preparing green food for them is eliminated and will prove a great economy.

Hens that have an abundance of alfalfa will lay eggs with very rich colored yolks and these eggs are usually fertile and produce healthy, vigorous offspring. An alfalfa range insures health, a good digestion and to growing chicks, a large frame. In buying a chicken ranch, one of the important questions is "will the land grow alfalfa?" Is there sufficient water to raise a good crop of alfalfa?

Alfalfa meal, or as it is sometimes called, Calfalfa, has been successfully used for hens. This is alfalfa hay ground up finely to form a meal. I have used this for several years and I find it sometimes good and sometimes bad. The analysis of it made by the University of California shows the protein content to be very high, and the nutritive ratio to be 1:3.3. This is the good meal. The poor meal contains too much fiber, and, as Prof. Rice of Cornell University remarked, "It was better for stuffing a bed than a hen." It all depends upon the quality of the alfalfa. Sometimes it is left until it is too old or is not properly cured, and is almost valueless; at other times it may have been dampened and become musty. When this is the case, it will disagree with the fowls and give them diarrhoea. To test it pour boiling water upon it and if it smells sweet, like hay, it is all right. If there is a musty, mildewy smell, discard it.

Preserving Eggs

Of twenty methods of preserving eggs tested in Germany, the three which proved the most effective were coating the eggs with vaseline, preserving them in lime water, and preserving them in water-glass. The conclusion was reached that the last was preferable, because varnishing the eggs with vaseline takes considerable time and treating them with the lime water may give them a disagreeable taste. These drawbacks are not to be found with eggs preserved in water-glass, which unquestionably is the best preservative yet discovered. The most difficult point probably in the use of water-glass for preserving eggs is its tendency to vary in quality. As a matter of fact there are two or three kinds of water-glass, and in addition to the fact that the buyer does not always have a distinct idea as to what he wants, the local druggist may not know all about it, or he may not know which kind is best for preservative purposes. The main use of these preparations for years has been the rendering of fabrics non-inflammable. This use in the Royal Theatre of Munich has rendered the place fire-proof by its use as a varnish in the fresco work, woodwork, scenery and curtains. It is also used for hardening stone and protecting it from the action of the weather. It was thus used many years ago, to arrest the decay of the stones in the British Houses of Parliament. The use of this medium for egg preservation is comparatively new, especially in this country, and it is not to be wondered at that dealers do not always supply just what is wanted.

Different Names for Water Glass

If we used the term soluble glass or "dissolved glass" in preference to either water-glass or silicate of soda, it might better describe just what we want, although one of the other names might be preferable when ordering of the druggist. This term expresses exactly what the material is. When we buy it by the pint or quart, we get dissolved glass. When we buy it dry, we get a soluble glass powder sometimes like powdered stone, sometimes white and glassy as to its particles. The powdered forms are supposed to dissolve in boiling water, but they do not dissolve readily, and must often be kept boiling for some hours.

Water-glass is made by melting together pure quartz and a caustic alkali, soda or potash, and sometimes a little charcoal.

Several of our Experiment Stations have made some rather exhaustive experiments with this dissolved glass in preserving eggs. The reports are, without exception, in favor of it. No other preservative is reported as being equal to this one. The stuff is invariably described as a thick or jelly-like liquid, and the proportions recommended are one pint of the silicate of soda to nine pints of water, although the Rhode Island Station reports experiments in which as low as two per cent of water-glass was used with favorable results. This as done to find out how little could be used, but this small proportion was not recommended. Further trials may show that less than nine to one may be reliable.

Directions for Use

The directions for use are: Use pure water which has been thoroughly boiled and cooled. To each nine quarts of this water add one quart of water-glass. Pack the eggs in the jar and pour the solution over them. The solution may be prepared, placed in the jar and fresh eggs added from time to time until the jar is filled, but care must be used to keep fully two inches of water-glass solution to cover the eggs. Keep the eggs in a cool place and the jar covered to prevent evaporation. A cool cellar is a good place in which to keep the eggs.

If the eggs be kept in a too warm place the silicate will be deposited and the eggs will not be properly protected. Do not wash the eggs before packing, for by so doing you will injure their keeping qualities. Probably by dissolving the mucilaginous coating on the outside of the shell. For packing use only perfectly fresh eggs, for eggs that have already become stale cannot be preserved by this or any other method, and one stale egg may spoil the whole batch.

I can speak from my own experience, for I have packed eggs in it for five years and shall do so again. We are fond of fresh eggs and use a great many, and I find it most convenient to have a jar or crock full of nice eggs always on hand. I have kept them myself for eight months and have no doubt but that I could have preserved them still longer had we not eaten them, for I found them to all appearances as fresh as if not over a week old. It costs about 1½ cents per dozen to preserve them.

The Kind of Vessels for Packing

Prof. Ladd, of the North Dakota Agricultural Station, spoke of receiving a few complaints that barrels were not proving satisfactory, the water-glass appearing to dissolve some product which deposited on the eggs. He thinks this might be attributed to the presence of glue, which had been used as sizing for the barrels. In such instances, charring the barrel inside with thorough washing thereafter, is recommended. Altogether the preference seems to be for glass or stoneware vessels.

Prof. Ladd's statement as to the satisfactory results of the water-glass method is very strong. He says: "This method has been tested in a commercial way, in nearly every state and part of our country, and we have not had to exceed eight adverse reprts." One of the stations affirms that the failures reported are probably due to receiving water-glass of poor quality.

It is also stated that these, like all preserved eggs, contain a little gas, and, when boiled, they will be likely to burst unless previously pricked through the shell at the large end.

As the entire processes of preservation are an effort to fence out germs, the recommendation not to wash off the mucilaginous coating which nature puts on the eggs, and also to use only boiled water, appear very logical. When we know just what we are aiming at, we are less likely to omit the little precautions which otherwise might seem like the whims of some mussy person. Too many people skip the essentials when trying to follow a formula.

I have kept the eggs in tin receptacles, five gallon kerosene oil cans, and large lard pails. These kept the eggs perfectly, but after a time the water and silica of soda rusted them in spots and the red rust formed a sediment on the eggs. This did not injure them as far as I could see, except giving them a brownish tinge, and on asking the druggist, he said he did not see why the tin should not be used, as the silicate of soda comes from the East in tin cans. If tin is used, it is best not to paint the cans or oil them, as the soda has an affinity for oil and will eat through it and the oil or grease may impart a disagreeable flavor to the eggs. Remember the eggs must be absolutely fresh, for one bad egg may spoil the whole quantity in the receptable.

Preserving in Lime

The process of keeping them in lime-water is as follows: Slack four pounds of lime, then add four pounds of salt; add eight gallons of water. Stir and leave to settle. The next day stir again. After the mixture has settled the second time, draw off the clear liquid. Take two ounces each of baking soda, cream of tartar, salt petre, and a little alum. Pulverise and mix; dissolve in two quarts of boiling water. Add this to the lime water. Put the eggs in a stone jar, small end down, one layer on top of another, and pour on the solution. Set the jar away in a cool place. This method is quite satisfactory, but not so good as the water-glass as the eggs are liable to taste of the lime.

White Wyandotte Cockerel

Capons

"Does Caponizing Pay?" We will consider the matter fully and from different points of view.

In Philadelphia and New York, in London and Paris, capons are considered a great delicacy, and as we in California become more metropolitan, capons will be more and more in demand. Eleven or twelve years ago when I had capons for sale I could not get more per pound for them than for the uncaponized fowls, as the Angelenos had not been educated in taste to the excellency of capon meat.

Capons are undoubtedly a more delicious dish at a year old than an uncaponized male bird of the same age. I had been led to suppose that a capon would be immensely heavier and larger than an uncaponized bird of the same age. This I found was not the case, the capons being rarely more than from half a pound to a pound heavier, if at all. My chief reason for caponizing was the desire to train capons for foster mothers of chicks. I wanted mothers that would not commence to lay as my hens did when the chickens were two or at most three weeks old and then desert them. In this I was thoroughly successful. The trained capon will mother chicks just as long as the chicks will stay with him, and after a little rest will take another brood and mother it again, clucking to the chicks, feeding them, defending them, hovering them better than the hen.

"Does caponizing pay?" Careful experiments have proved that the increase in weight is by no means so great as the public has been led to believe. It takes capons at least a month to sufficiently recover from the operation to catch up with their former mates in size and when they come to a marketable age they seldom weigh a pound more than the uncaponized birds of the same breed and age. The gain, however, in price is in their favor for it about doubles that of the other. This sounds like a strong argument on the side of the capon, but again the cost of production is an essential factor in the study of the question. It will cost as much to produce a ten pound capon as to produce three or four young chicks of the same combined weight; in fact with food at the present price I really think it will cost more.

"Does caponizing pay?" I knew a lady about three years ago who sold four capons for sixteen dollars. She was so much encouraged by this, for they averaged 38 cents a pound, that the following season she drove around the country buying up little cockerels and caponizing them. She was very successful in operating, rarely losing any, but as she only stayed in the business one year, I think she did not consider it very remunerative.

Easy to Learn

The art of caponizing is simple and easy to learn. In France the farmers' wives and daughters have done the caponizing for centuries and practically without instruments except a sharp knife. In this country and age, we can buy a case of the best instruments, with full instructions for use, at low cost, and the Agricultural stations of some states give free demonstration lessons to anyone within the state. The Rhode Island College gives lessons in caponizing in connection with its poultry course and also sends out, free, a book of instructions. By following these instructions and experimenting for the first tme on a dead chicken, any one that is deft can learn it. The operation is performed with apparently little pain to the subject and the minute the bird is released it will eat heartily and walk around as if nothing had occurred.

In foreign countries the art of caponizing has been known and practiced for ages, yet it is not so common nor are capons so plentiful but that prices rule high and capons are considered the choicest of viands and above the reach of any except the rich. In this blessed country there is no reason why the producers of poultry should not feast upon capons, besides having the satisfaction of producing and marketing strictly high class poultry.

Favorite Breeds for Capons

In New England the favorite breeds for caponizing are the light Brahmas and the Cochin and Brahma crosses. They are chosen on account of their large size and slow growth to maturity. The Plymouth Rocks follow, together with the Orpingtons and Wyandottes. The smaller breeds make, of course, much smaller capons, still they are popular in small families where large size is not required. I have personally caponized only my White Plymouth Rocks. Nothing could be better than capons of this breed. At nine or ten months of age they are in their prime and the juciness and flavor of their flesh is superb.

Among the advantages of caponizing are, the birds may be kept together in large numbers, will not quarrel or fight, will not harass the hens and pullets, will not misuse the little chicks, bear crowding and take on flesh more rapidly than cockerels. They make when trained most excellent mothers for little chickens, sheltering them under their long feathers and great wings.

Best Time for Caponizing

The best time for caponizing is in the early fall, for the reason that the heat of summer does not then retard recovery and also because the late (June hatched) cockerels are then of the best size.

The best size is from two and a half to three pounds weight and this would be about the weight of June hatched chickens of the American breeds which if caponized in September will be well grown and in good shape for marketing in March, the time of the highest prices.

It is to the farmers, however, that the recommendation to caponize their cockerels for the family table should appeal most strongly, for they are the class that would be most benefited by having good capons to eat. It is a simple task to caponize forty or fifty birds and by that simple method a farmer can provide his family with dinners which will be the envy of his less fortunate friends.

The question "Does caponizing pay" may be answered, "Sometimes it does and sometimes it does not." ·

Counting the cost of the food for at least six or eight months, the cost of the operation, unless one can do it oneself, and the time and care during that time, also the space necessary which might be better occupied by pullets, we are apt to think that it does not pay.

But again considering the high price obtainable, the peace and quietness of the poultry yard, and the excellence of the flesh, the ease with which the birds are fattened, and the growing demand —does it not pay?

It is almost impossible to keep together any number of male birds in their natural state after they reach a certain age, without a deterioration in the quality of the flesh, the color of the skin and the general market value of the birds as poultry. Caponizing, by doing away with all strife and worry, gives them a chance to turn the food consumed to flesh instead of expending it in the wear and tear of ceaseless fretting and action.

There is a best time for operating, a best age and a best size. I have found the best age for caponizing to be about three months of age for the American breeds; after the age of four months most young males of these breeds cease rapid flesh development, and the food they consume, that might be made into positive profit, results in little more than frame and ambition to whip some other bird; while if they are caponized at the right age it makes it possible to keep them a sufficient length of time so that one can realize a gain from the work that goes into the hatching, feeding and care of the first four months of their life. Capons do not grow larger than cockerels until they are about six months of age. After this there comes a change in the capons, their combs do not develop, but their hackle and tail feathers grow long and beautiful.

As a rule it is the later hatched chickens that should be caponized for the spring market. We all know that the markets are flooded with chickens in the late summer and fall and that the low prices of poultry at that time return the raiser but very little profit. These somewhat late chickens if caponized in September will grow to a good size and be in the best market condition for the spring season when the turkey market is on the wane, and this will bring the grower a handsome profit.

Capons as Brooders

Capons make excellent mothers when trained to it. Some breeds would probably make more affectionate and attentive foster mothers than others. I can personally answer for the Cornish Indian Games and Plymouth Rocks. I have also seen beautiful Brown Leghorn capons that had raised several broods of chickens. Cockerels hatched in November, December and January, make excellent capons for brooding. They should be caponized at about three months of age. Should be gently handled and never frightened, when they will become perfectly tame. The capon with its changed nature is even more timid than a hen or pullet, and for this reason should be separated from any of the older fowls and kindly treated.

Capons should be trained at the age of about six months. They are easier to train at this age than at any other time, generally, but I have trained them at ten months of age. To train them, I keep the bird in solitary confinement for a few days, placing him in a cracker box; place water, grit and sand in the box the same as though preparing for a hen and her brood. After two or three solitary nights and days I put two little chicks under him at night; they snuggle up under him, and he is quite glad to have the little fellows for company. The next morning he will look a little surprised perhaps, but usually takes them immediately, and soon begins to cluck to them like an old hen. The following evening I put as many as I intend him to care for under him, and before going to bed at night, see that all the little fellows are under his sheltering feathers. My object in using a cracker box is that it is about the proper height to make it uncomfortable for the capon to stand upright and he will sit for comfort; the little chicks get closer and make friends quicker, and have an opportunity to nestle under the capon as they would a hen. This training should be done in pleasant weather, because the chicks will not be hovered at first as well by the capon as the hen, and I use only a few chicks the first time, because a young capon with his first brood does not hover them like a trained one.

The Whiskey Treatment

Hen hatched chicks take to a capon without any trouble, but chicks which have been several days in a brooder seem afraid of the capon, and instead of running to him to be hovered, huddle in a corner, so it is best to put them straight from the incubator under the capon. A writer on this subject says: "Should one of the capons pick the chicks I would take him out of the box and swing him around in a verticle circle at arms' length until he was sick, then put him back again. If he attempts the same thing again, I take a small glass syringe and inject about one tablespoonful of good whiskey into his crop through his mouth, and after this treatment he is pretty sure to take to the chicks. He becomes so docile that he allows the chicks to pick at his face and will not pick back at them. When you notice this, you can rest assured that he is on the right road."

I have never tried the whiskey treatment, and have never had any difficulty in training a capon. Capons have proved far superior to hens in brooding chicks, in fact they excel all other methods, either natural or artificial. The hen, especially "bred-to-lay" strain, deserts her brood at too early an age, and some hens, especially the pullets, with a first brood, are often very stupid at caring for them. I have known a pullet to hover her chicks in a thunder storm in a gully where the water rushed until they were nearly all drowned. Pullets do not seem to have sense enough to "come in out of the rain," while a good capon, when once he has been taught his way home, will bring the little ones to shelter without any trouble. The capon will defend his little brood most vigorously against cats, dogs or any animal. He seems to develop all the latent parental affection and lavishes it on his young charges as if his one and only object in life was to care for them.

When Changing Broods

When the chicks are old enough to take care of themselves, before entrusting another brood to his care, he should have a rest of at least two weeks, especially if the next brood is to be of another color. During the two weeks rest he will forget the color of the chicks he had and will not be so apt to object to the new ones. We all know that hens will sometimes object to chicks of a different color and will oftentimes kill them. When once trained, a capon is very little trouble and will care for brood after brood without any more training than I have mentioned. Capons can be kept over several seasons. I have heard of some being used for eight years, but mine were usually fattened and made a toothsome dish after two years' service.

It is not difficult to learn how to caponize. The tools or instruments necessary are to be found at the poultry supply houses. The price for a set of instruments is from $2.50 to about $4.00, largely depending upon the case in which they are contained. The poultry supply houses have books of instruction for caponizing, and at some of them you can learn the name of persons who, for a small sum, will caponize for others. It would be a good plan for several neighbors to join together and have the person caponize 50 or 100 in the same day. In this way, it would make the price lower.

Capons are not much larger than cockerels of the same breed and age. The difference is in the table quality of the flesh. It is juicier and more tender, just as steer beef is superior to any other beef.

Black Orpington Cockerel

DRYING BOX

A box is often used like the picture for drying fowls after their bath. It is heated by a lamp and current of hot air; is considered very effective and enables one to operate in a cooler room—a great advantage to the operator.

Ready for the Show

Getting Ready for the Show

To wash and dry four fowls is a good day's work for anyone; even an expert. I have washed fowls for every show at which I have exhibited; the last time a helper and I washed thirty-two, so I speak from actual experience. In washing the fowls I used four tubs instead of three; the first tub was the clothes boiler because I could make the suds deeper in that than in a tub. The suds must be made with ivory soap, or white castile, and be as warm as one can bear one's hand in. I used a good soft flannel wash cloth and holding the fowl's shanks between the fingers of my left hand, immerse it, wetting it thoroughly to the skin, using the flannel to scrub and wash it most thoroughly till every feather was perfectly clean; it takes about half an hour in this first tub and two persons are better than one, as every feather should be gone over carefully and should look pure white. When the bird is clean, lift it out, and whilst one holds it let the other gently pass the hands down the back to squeeze the suds out. Then immerse in the second tub of warm water and with a clean sponge wash and rinse it again; taking it out, squeeze the water out, put it into tub three, in which the water is a little cooler, and finally when all the soap is thoroughly rinsed out, put the bird into the last tub, in which there is cold water with a little blueing in it, as much as one would use for clothes. If the soap has not been rinsed out the blueing will stick in places and the fowl will look worse than if it had been left dirty.

Taking the fowl out of the last water, I ran my hand gently down its back to slightly squeeze the water out, then wiped it with dry linen towel, and then made it flap its wings by throwing it in the air whilst holding its legs. The room has to be very warm—110 degrees Fahrenheit, and must be kept at that heat until the fowls are dry. If they are allowed to dry in a cool room the feathers stick together, but in a hot room they become fluffy and beautiful. The fowls must be washed in the morning, for it takes several hours for them to dry. After they are washed, I give them a quarter of a teaspoonful of red pepper mixed with a little butter to keep them from taking cold. I put them in a coop with clean fresh white straw to dry them, and feed them dry grain. They should have grain as their principal diet for two weeks before the show.

Cleaning the Shanks

The afternoon after having bathed the fowls in the morning, I commence with the first one that is dry. I first have a sharp pen knife, sharpen half a dozen matches to a point, spread a clean white cloth on my lap, sit down in my rocking chair near the kitchen window, my helper hands me the fowl, and I wrap the cloth round it, to prevent its flapping its wings. I then take a pointed match and, commencing at one toe, I clean under each scale, just as one cleans one's finger nails, and gently, but firmly clean each one. As soon as the point gets worn off, I take another match. I have tried tooth picks, and finger nail cleaners, but found the matches the most satisfactory for the fowls and always handy. It takes from an hour to an hour and a half to clean every scale on a fowl's shanks and toes. After they are clean I either wash them in warm soap suds and wipe off without rinsing, or rub them over with a raw carrot (some say a potato will do but I have not tried it). This is to slightly moisten the scales, for by handling for over an hour they become hot and dry, and lose their luster. I know a man who every day for a week rubbed his cockerel's shanks with a carrot, and they shone as if newly varnished, and were a beautiful yellow. Moisture keeps the shanks yellow, but the adobe and alkali on some farms bleaches and ruins them. For those unfortunate enough to be on adobe soil, I would advise them to wash and clean the fowls' legs a month or two before the show and grease them slightly with olive oil, lard or butter, rubbing it well in and wiping off. Do this every week; this will counteract the bad effects of the adobe soil, and your fowls will come out all right. "Lots of trouble," you say. Yes! but it pays. If you don't use every effort to win, some other person will, and the winner deserves to win.

Without Washing

Now I will let you into a secret. This washing of the fowls, when well done (and it must be well done or not at all) makes them look beautiful, there is no doubt about that, but it is very trying upon the person who does it; many a bad cold have I taken from being overheated, spending four hours of a morning in a room with temperature at 110 degrees Fahrenheit. It is also very trying on the fowls, and they sometimes take cold; for this reason, as a preventive I administer the red pepper.

The secret is a way of cleaning the fowls without washing them:

Give them a large scratching pen under shelter, or in the house, keep this filled twelve inches deep in dry, clean wheat hay for a month before the show, adding more wheat hay every few days. Feed them all the grain in this. They will almost bury themselves in the straw, and it will clean them wonderfully, the straw sliding down their backs keeps up a continual wiping down and polishing off, and at the end of a month they are glossy and clean. You must be very extravagant with the wheat straw or hay, frequently renewing or adding to it so as to have it clean and deep. In this way you can make the fowls clean themselves. Of course the feet and legs must be washed and scrubbed with a toothbrush and good warm suds, and then cleaned as I have described with a pointed match under each scale.

Turkeys and How to Raise Them

Turkeys have been called the "farmers' friend," and there is no doubt that turkey raising on a small scale is more profitable than any other branch of the poultry industry and that turkeys will bring larger cash returns than any other stock upon the farm. They cost very little to raise, they eat the waste grain in the fields and barnyard, besides the seed of many harmful weeds. They consume an immense number of grasshoppers, grubs, worms and insects which would otherwise greatly injure the farmers' crops, and they are not difficult to raise if they are not overfed.

One writer asks if chick feed is a proper and safe food for little

A Magnificent White Holland Tom

turkeys, and another requests me to tell her exactly how I feed and care for the little turkeys.

Chick food is neither a safe nor a proper food for little turkeys, although it is a most excellent food for little chicks. In fact, you may be sure of success when you feed it to chickens and failure if you feed it to turkeys. Later on I will try to explain this.

Now, as to my way of rearing turkeys. I am glad to give it because now I raise every turkey that is hatched barring accidents, as some will drown in the cows' trough and occasionally one or two get stepped on, or the door blows on one or the puppy worries another. None die from disease.

I do not pretend to say that mine is the only way but I do say that not only do I succeed in raising turkeys, but those who have followed my directions were as successful as I have been, and those that met with failure did not follow my plans. I have been criticised as too fussy and particular about little details, but I think it pays to take good care of the little things for a few weeks, for turkeys are delicate only when they are little, and if properly cared for then will be strong and hardy when they mature.

Grandmother's Recipe

At my grandmother's the recipe for feeding little turkeys was as follows: "Leave them in the nest twenty-four hours or until the mother turkey brings them off; then give them only coarse sand and water to drink. Meanwhile put some fresh eggs in cold water to boil; let them boil for half an hour; then chop them up, egg-shells and all, quite fine; add an equal amount of dry bread crumbs, and always, always, some green food chopped up finely."

Lettuce, dandelion or dock were the green foods at grandmother's and the explanation given me was that if they are fed without having green at every meal they soon become constipated, then get sick and die. The secret of her success was the tender green food and the grit, a pinch of coarse sand being sprinkled over the food of each meal. As the little turkeys grew, a little cracked wheat and later whole wheat was added to their food. That was the only grain given. This was grandmother's recipe for raising turkeys.

The way I feed and have fed for years is as follows: When the little turkeys are twenty-four hours old I put freshly-laid eggs into cold water and boil them for half an hour; chop them up fine, shell and all; add equal parts of bread crumbs; feed dry, taking away what they leave, feeding the mother separately.

The next day I feed the same, adding very finely chopped lettuce or dandelion leaves or green young mustard leaves and tender young onion tops. This is their breakfast and supper. For dinner they have a little curd made from clabber milk, cottage cheese some call it. In a few days I add cracked or whole wheat to their supper and if I am short of bread crumbs I add rolled breakfast oats to the egg and bread crumbs. I always chop up an onion a day with the egg and bread crumbs unless the onions tops are very young and tender. Onions are an excellent tonic for the liver and kidneys and prevent worms and cure colds; so I use onions freely both for turkeys and chickens. In a few days I commence to add wheat to their food and at two weeks of age I gradually arrive at giving them wheat and rolled oats for breakfast; in the middle of the forenoon a head of lettuce to tear up and eat; at noon cottage cheese, and about four or five o'clock their supper of egg, bread crumbs or rolled oats, lettuce and always the chopped up onion.

I give them clean water three times a day in a drinking fountain or if I have not a fountain I make one out of a tomato can. Make a nail hole in the can about half an inch from the top, then fill the can up to the hole with water, invert a saucer over it, and holding the saucer tightly to it, turn it over quickly. This makes a good fountain for the water will come slowly out of the nail hole into the saucer. I give the turkeys a similar fountain of skim milk, also. A word about the cottage cheese. I am very particular in making it not to allow the clabber milk to become hot. I use either a thermometer, letting the heat only come to 98 degrees, or I keep my finger in the milk and as soon as it feels pleasantly warm I take the milk off the fire, pour the curd into a cheese cloth bag and leave it to drain. If the milk scalds or boils the curd will be tough, hard like rubber and indigestible enough to kill turkeys or chickens.

Overfed Little Ones

When I lived in the home of the wild turkey, Oklahoma and Kansas, I learned much about the care of tame turkeys. There "corn is king" but I was cautioned never to give corn to the young turkeys until after they "sport the red." That is until their heads and wattles become red, which happens at about three months of age. For it was said that corn always sours on their stomachs. It was there I heard of a man who brought up his turkeys on nothing but onion tops, curd and grit, and they did well.

One of my experiences in the land of the wild turkey may serve as a warning to others. I had a good old Buff Cochin hen who was mothering a brood of nice little turkeys. She was most assiduous in her care of them; she clucked to them all day; called them up to eat all the time and it was surprising to see how those little fellows grew, when one after another they began to droop and die, till only one little fellow was left. The other turkeys under turkey mothers were doing well, so I took the lone little one one night and put him under a mother turkey out in the meadow and saved his life. The old hen had over-fed the others. Chicken hens are too anxious to feed the little turkeys. They scratch for them, coax them to eat, and the little turkeys are such greedy, voracious little things that they over-eat and in consequence die. I much prefer to bring up little turkeys under a turkey hen or even in a brooder, rather than under a chicken hen. The best way of managing a hen is to keep her in a coop letting the little turkeys run outside or else tie the hen under a tree by her leg. I only feed the little poults three times a day just what they will eat up clean in ten minutes. With a turkey hen I can leave wheat in a trough always accessible and she will never over-feed the young. The turkey mother will take a few mouthsful herself and then move slowly and deliberately away and her babies will follow her, having only taken one or two grains each. This is more like the nature of the wild turkey and the nearer to nature one can keep in raising turkeys the better will be our success.

Nature's wild turkeys are only hatched in the spring when there are grubs and worms in abundance, with plenty of green grass and tender leaves and no grain but what is sprouting, and above all Nature never mixes mashes to turn sour and ferment on the little stomachs. The hard-boiled egg and the curd takes the place of the bugs and the grubs, for we cannot supply the turkeys with anything like the amount of grasshoppers, grubs, worms, larvae of insects which Nature provides in the haunts of the wild turkeys. Another lesson we may learn from Nature's book: Wild turkeys are only to be found where there are springs and streams of pure water and they never wander away from the water. Give the young turkeys plenty of clean, pure water to drink.

There are two chief causes of mortality in little turkeys—lice and over feeding. Before giving the little turkeys to the mother to care for, dust them well with "buhach," and continue to do this once a week until they are too large to handle. Look for lice on the head and on the quill feathers of the wing and rub the powder well into them. Lice and over-feeding kill thousands of little turkeys . Overfeeding kills more than lice and if it does not kill them it stunts their growth, and unfortunately until they begin to die at about six weeks of age, one scarcely realizes that they have been overfed.

Little turkeys have voracious appetites, and if allowed to do so will eat too much and it only takes a few weeks for them to eat themselves into their graves. If they hunt for their food as the wild turkeys do they take it leisurely, just what they can easily digest, exercising between each mouthful and just enough is digested and goes into the circulation to keep them healthy. I never feed little turkeys all they want, only what they need and I always keep them a little hungry.

Keep Liver Healthy

I can tell you just how overfed turkeys will die. First they will walk slowly, lagging behind the others as if tired, then their wings will droop and they will look sleepy and will not eat, will look at the food as if they wanted it, but were too lazy to pick it up, then diarrhoea will set in, the dropping will become yellow and sometimes green, and death will soon follow. If you hold a postmortem examination, as you should do over everything that dies in the chicken yard, you will find the liver of these little turkeys has yellow or white spots on it, and on cutting into it you may find that these spots are small ulcers that extend through it. Sometimes these ulcers are quite offensive. This comes from over feeding which gives the liver more work than it can do and it breaks down.

The liver is the largest organ in the turkey's body and it seems to be the most delicate. If you can keep that healthy you will have healthy turkeys. Onion and dandelion leaves are tonics for the liver and the green food keeps it healthy, whilst the animal food and a small amount of cereal will make the frame of the turkey.

But suppose you should see one little turkey in the brood beginning to walk slowly, what should you do? I will tell you what I would do. I would catch that little turkey and give a Carter's Little Liver Pill and follow this the next day with a little Epsom salts for the whole flock and cut off some of the grain in the feed. You will probably save the flock, but they will be stunted in their growth, and their liver many months later, may break down from their being weakened by that first attack of liver trouble.

Chick Feed for Turkeys

Now, about the chick feed. It is composed of a number of different grains. Some of these grains are extremely difficult of digestion for turkeys. The chief of these are cracked corn, Kaffir corn, Eygptian corn, sorgum seed, millet, etc. I could scarcely believe this until I had ocular demonstration of it. Then I discovered that cracked corn did not commence to digest in the crop, as the gastric juice of the crop does not seem to have any influence on it. It passes through the crop and on through the procentriculus to the gizzard, arriving there hard and not in the least softened or digested and there it commences to ferment, causing diarrhoea or else passing away without digesting. I am not scientific enough to know the reason of this nor why wheat should be softened in the crop and partly digested before reaching the gizzard, but I know that it is so. They told me in Kansas that corn soured on the turkeys' stomachs, but it does not exactly sour, it ferments—and there is where the trouble comes in.

Sour milk is sour, but this is from lactic acid, and lactic acid seems beneficial to turkeys, whilst the souring of grains, bran, cereals of any kind, bread or corn meal is a ferment, and ferments are very injurious to fowls of all kinds and especially so to turkeys.

Mrs. Charles Jones, the best authority on turkeys in the United States, agrees with me about feeding turkeys. She writes:

"A diet of part corn agrees with chickens, but I have never yet fed corn in any form to young turkeys but that sooner or later they would give up the unequal contest. A little neighbor girl that had a great deal of the care of turkeys said the least little bit of corn meal makes them die. She had learned this by watching them as she fed them."

1100 Gleaning Wheat

It was my privilege to visit a turkey ranch in the San Joaquin Valley some time ago and what I say there made me wonder that there are so few large turkey ranches in California.

There were over 1100 beautiful turkeys gleaning the wheat over many acres of stubble. These great turkeys had been hatched near the barn in shed-like coops, under turkey hens. They were kept in the yard until about five or six weeks old, when they were driven out with their mothers upon the wheat stubble to rustle for their living, to pick up the wheat that would otherwise be lost. All these turkeys roosted in the open air and to this and the simple life, working for and finding their own living, may be attributed their healthiness.

There are many beautiful valleys in California where turkeys may be grown to great advantage by the hundreds and even thousands, but even on small ranches a few may be kept.

MORE ABOUT TURKEYS

There is no need for any sickness amongst turkeys whatever in California, if they are properly cared for, and I think eventually California will supply the Eastern States with their Thanksgiving and Christmas dinners, for they have there a disease among turkeys which is so serious that it is decimating, and, in some places, wiping out whole flocks of turkeys. The disease is called "Blackhead," as the head in some instances turns black or dark colored before or at the time of death.

The Oregon Experiment Station has recently issued Bulletin No. 95, by E. F. Pernot, on Disease of Turkeys. This bulletin contains information of very great importance to the turkey raisers of the State. It treats the subject of Blackhead, explaining the cause of this disease, the symptoms, and treatment. This bulletin, which may be obtained free on application to the Experiment Station, Corvallis, Oregon, should be in the hands of every turkey breeder in the State.

In sections of the East, Blackhead has almost wiped out the turkeys, and the same thing is liable to happen in this State if proper measures are not taken to prevent it.

I give here a brief summary of Prof. Pernot's bulletin:

Symptoms:—Diarrhoea is the most pronounced symptom. The discharges are frequent, thin, watery, and generally of a yellowish color. This, however, sometimes occurs from other intestinal disorders and does not alone signify the presence of the malady. The next symptom is the drooping tail, followed by a drooping of the wings after which death soon ensues. When the disease is at its height, the head assumes a dark color, hence the name, Blackhead. Young turkeys are much more susceptible or they may be more delicate, and cannot withstand the invasion of the parasites so well. They begin by moping and hunching up as though they were cold, diarrhoea soon sets in, the tails droop, then the wings droop, and they go about uttering a pitiful "peep," after which they soon die. A blackening of the head does not always occur.

It is only by careful post mortem that the true cause of the disease may be determined.

The Cause:—The disease is caused by animal parasites, which can be detected only by the aid of a microscope. Because of their minuteness and growth in the mucous membranes of the digestive tract, they are easily carried by the excreta to food, which upon becoming contaminated, transmits them to other fowls. This is the usual means of infection.

Remedies:—Food given to fowls should never come in contact

with their droppings, as one bird with the disease will infect the feeding ground of others. Better sacrifice the bird at once than run the risk of spreading the infection to the whole flock. A sick bird should be removed from the flock and placed in close quarters, which may afterwards be disinfected, or the bird may be killed at once and then should be burned. Medical treatment is not very successful, owing to the difficulty of reaching the parasites at the seat of the disease; yet treating them with some of the following remedies is well worth the trouble: Sulphur, 5 grains; sulphate of iron, 1 grain; sulphate of quinine, 1 grain. Place this amount in capsules and administer one night and morning to each turkey for a week. If the bird does not respond to treatment, kill it at once without drawing blood, and then burn the carcass, disinfecting the coop.

A solution of carbolic acid prepared by mixing five parts of the acid to 100 parts of water makes a good disinfecting solution, or chloride of lime, 5 ounces to 1 gallon of water, is good. Corrosive sublimate in the strength of 1 ounce to eight gallons of water, is a strong disinfectant, and may be used with a broom or spray to wet every part of the coop and floor, but it is poisonous and must be handled with great care. To disinfect the entire premises when the fowls are running at large is impracticable; but lime should be used freely on the droppings beneath where they roost. When the disease becomes seriously destructive it is more than likely all the flock are affected, and it may be necessary to destroy all the remaining birds and disinfect the premises as thoroughly as possible. In such cases it would be better to suspend the raising of turkeys for one year.

Liver Complaint

Personally, I have only met once with a case in California which might be called Blackhead. I have seen many cases of common liver complaint, and by my directions others have succeeded in curing many of these.

Dr. Salmon tells us that the seat of the disease called Blackhead is in the caecca. The caecca is sometimes called the blind bowel; it is a sort of "appendix" in the turkey, having no outlet. It is two lobes of bowel united by a ribbon of fat (the pancreas). In Blackhead and also in some cases of liver complaint, an abscess forms in one or both caecca, but this can only be discovered after death, and I have only found it in a postmortem of one turkey. The fact is, I have been so very "lucky" in raising turkeys that now I rarely even see a sick turkey, and I have many letters from our readers telling me they have cured their turkeys by my directions, so I will repeat them again for the benefit of new comers.

First, liver complaint comes from wrong feeding, or over-feeding, which has overworked the liver; secondly, Blackhead comes from a parasite; thirdly, the symptoms of both diseases are almost exactly the same in the first stages. Dr. Cushman, in discussing this matter, decided that when the bright yellow diarrhoea comes on showing liver trouble, the remedy is "something bitter and something sour." This is easy to remember. He also recommends no food but green food and says that turkeys have been known to cure themselves by living on acorns.

My remedy is first a liver pill followed by quinine for a week, and sour milk and no food but onions and green alfalfa or grass, keeping this up until cured.

I have a letter from a successful turkey raiser of Long Beach, near Los Angeles. She writes: "I wish to tell you my experience with liver sick turkeys. I had a gobbler weighing eighteen or twenty pounds, and I made the mistake so many do of allowing turkeys and chickens to run together; my experience is that turkeys, especially toms, will not stand such quanitites of food that hens do. Well, he got very sick, so bad he was as light as a feather, and my cure, which never fails, was administered—a bottle of Jamaica ginger and a bottle of liquozone were procured. I put him in a clean, large coop and he lay on a bed of straw for days, so weak he could not stand. The first day I gave him one teaspoonful of the ginger and one teaspoonful of the liquozone mixed and diluted until it was not too strong, giving two or three spoons every hour of the diluted. The next day giving it three times a day; after that twice a day. I did not allow him anything to eat, but of an evening gave him the smallest sized capsule of quinine. Kept that up until he began to get good and hungry, then fed him a few grains of wheat, only about six grains, and a little speck of alfalfa. I have found that feed kills them every time when they are so sick. I never failed to cure the worst cases if I treat them like I tell you. Then if they hump up again and begin to get sick again, I give them a dose in the evening. The ginger warms them up and starts circulation, and the liquozone kills the germs."

Liquozone is very acid, it tastes like sulphuric acid and water, and I have no doubt that my friend's cure is a good one. Remember Dr. Cushman says "something bitter and something sour," and if your turkeys get sick, try it immediately.

A Bronze "Gobbler"

Ducks in Their Varieties

In the springtime of the year in the East the big duck ranches hatch ducks by the hundreds of thousands, but in California, or at least in the neighborhood of Los Angeles, there are not such large ranches, and ducks do not seem as popular. Probably some farmers have had a few in their yard at some time, just to give them a trial, and have found them a continual nuisance, as they greedily eat the whole allowance of food from the expectant chickens and dabble in their drinking vessels, so they have to be continually cleaned and replenished, and with great injustice to the ducks they have let this prejudice them, where if they had kept the ducks separate, they would have found them easier to raise than chickens.

Ducks grow faster and are ready for the market earlier than chickens; they are not troubled by the diseases of hens, neither do they have lice, except if raised under a hen when very young, before the feathers grow, the gray head-lice may get on their heads, crawl into their ears and kill them, but this is before they feather out. Mosquitoes which are very troublesome in some places to the chickens, causing great mortality, never trouble ducks, neither do fleas or ticks. I think the reason for their immunity from vermin is that their feathers are very oily and thick and the down under the feathers is an extra protection. Hens require a dust bath, while ducks require a water bath to keep them clean and healthy.

Most of the popular varieties of ducks can be raised and bred without water to swim in, but on the very large duck ranches a supply of running water so that they may have fresh water to drink, and a bathing place for the breeding ducks, is a great advantage.

Ducks should be kept entirely away form chickens and turkeys, as they pollute water so badly it makes the other fowls sick. I found on my small ranch where there was only water piped in, after trying various plans for watering the ducks, an easy and convenient way. I had a barrel sawed in two, two-thirds and one-third. I knocked the head out of the larger end and buried that part, making it deep enough so the top of the barrel was just below

Some Fine Specimens of Indian Runners

the ground; any box with no bottom would do as well. The one-third of the barrel had a bunghole in the bottom. This one-third barrel I placed over the sunken one. I had a broom handle which fitted into the bunghole and every day I let the dirty water run through it into the bottomless barrel and it soaked away. In this manner I gave my ducks fresh water and a clean bath every day. I found if I sawed the barrel exactly in half, it made the top part deeper than I wanted, and the bottom not deep enough.

The Varieties

I have successfully bred the following most popular breeds of ducks and think a slight review of them may be interesting and helpful to beginners: The Aylesbury, Pekin, Indian Runner, Buff Orpington Duck and the Muscovy.

The Aylesbury

The Aylesbury, called after a town in Buckingham, England, are about a pound heavier than the Pekin. The standard weights being, drake, 9 lbs.; duck, 8 lbs.; young drake, 8 lbs.; young duck, 7 lbs. Their color is pure white, with pinkish-white beak and shanks. They are extremely popular in England and are hardy and vigorous. There are not many breeders of them in this country, but an Englishman, Mr. V. G. Huntley of Petaluma, who has imported some exceedingly fine Aylesbury ducks from England, says he has a large demand for them, as they are a rarity in this country. He considers their flesh better than that of any other variety of ducks. In plumage the Aylesbury are a pure spotless white with hard, close feathers that glisten in the sunlight like satin. The advantages claimed for this breed are the easiness with which it is acclimated, its early maturing, its great hardiness, its large size, being heavier than any except the Rouen, its great prolificancy and its beauty.

The Pekin

The Pekin is undoubtedly the most popular breed on the large duck ranches in the East, where thousands of them are fattened and turned off every season. This breed is variously called the Imperial Pekin and the Mammoth Pekin and Rankin's Pekin. It was brought to this country from China in the early seventies and immediately took the first place as the most prolific and rapidly growing duck on the market. In shape and carriage the Pekin has a distinct type of its own, which by some is described as resembling an Indian canoe, from the keel-like shape and the turned-up tail. Though Pekin ducks may not merit all that is claimed for them by enthusiastic breeders, it is certain that the duck business could not have attained its present proportion without the Pekin duck, and that as a market duck this breed takes the lead. They are hardy, quick growers, thrive in close confinement and are ready to market at ten weeks of age. The plumage is soft, more downy than that of other varieties and is of a creamy white in color. The beak is of a deep orange yellow, and, according to Standard, should be free from black marks. The shanks and toes are reddish orange color.

All ducks are of a timid disposition, and the Pekin more so than those of other breeds; in fact, they will injure themsevles so badly if frightened by cat, dog or a stranger, or by being caught up, that they may have to be killed. A fright, if not fatal, will take off several days' growth of the young, and stop the laying of the adult ducks.

The Rouen

The Rouen duck, so named for a city in Normandy, where they are supposed to have originated, are still bred there in large numbers. The Rouen duck is a fine market bird, but does not mature as early as the Pekin or Aylesbury. It is easily fattened, hardy and quiet in disposition and not as nervous as the Pekin.

The Rouen drake is a magnificent colored bird. Neck and head are irridescent green, breast wine color and the lower part of the body delicate steel gray, penciled with very fine black lines. About June a remarkable change takes place in the drake. He begins to lose his lustrous feathers, those of the neck dropping out, being replaced by feathers of a russet brown. The magnificently colored drake is clothed in sober hues for the summer. In October he again resumes his gorgeous raiment.

The Indian Runner

Many years ago Indian Runners were brought from India to England by a sea captain, hence the name "Indian," while the "Runners" came from their great agility. They do not waddle like other ducks, but run more like a plover, and are very quick in their movements. In England their good qualities quickly captivated the thrifty farmers. Individual ducks there have made a record of 225 eggs per annum. Here in California I had ten ducks which laid 2331 eggs in one year. I think the climate of California more nearly resembles that of their native land and their laying is never checked by cold or snow, so that here they lay better than in England or the Eastern States. In India they were bred for their laying and table qualities, no attention being paid to the color of their plumage; all the Indians cared for was the eggs, and they layed eggs galore. English breeders claim that eight-year-old ducks of this breed will lay as well as yearlings, and on this account, and their capacity for forging, they have become very popular in England and Australia.

While the weight of the matured Pekin is greater than that of the Indian Runner there is more meat in proportion to their weight in the Runners on account of the smallness of the bones; the meat is also of a much finer quality, finely grained and juicy and resembling in flavor the much extolled canvas-back duck. The eggs of the Indian Runner are an ivy white in color, greatly resembling Minorca eggs, very delicate in taste, and in England their eggs are in great demand in the tuberculosis sanitariums on account of their delicate flavor, richness and nutritive value, and absolute freedom from tuberculosis taint, and there is a higher price paid for them than the hen's eggs.

The standard color of the Indian Runners in this country is fawn and white. In England they also have the black and white, the brown and white and the pure white.

The Buff Orpington

Buff Orpington ducks are a breed of Mr. William Cook's making. He named them as he did the Orpington hens, after his own place in Kent, England. The color of the Buff Orpingtons is a soft shade of buff, the drakes having rich brown heads. The Buff Orpington has a good deal of the Indian Runner blood in it, and from this source its laying qualities are gathered. Mr. Cook claims they are better layers than any other of the duck family. Many of them lay a beautiful green egg, although a greenish-white is the usual color. These ducks weigh a pound and a half more than the Indian Runner, are large and more plump birds, maturing early, and one of the best market birds.

The Muscovy

The Muscovy Duck is not largely bred in this country. They are not like any other ducks and do not interbreed with others. It it a native of South America, where it may still be found in its wild state. It comes in two varieties, white and black and white. The males are much larger than the females. I had one weighing fourteen pounds. Both sexes have cruncles at the base of the beak; these become larger every year, giving them a vulture-like appearance. Muscovy ducks are rather awkward in the water, preferring to live on the land. They are pugnacious and ill-tempered, and, although they have web-feet, they have very sharp claws that can, and do, scratch in a most unpleasant way. They are strong on the wing, flying easily over the barn, and they like to perch on the roof. They are good setters, and their eggs take thirty-five days to incubate.

Hatching and Brooding

The first thing the amateur needs is first-class breeding stock or eggs of the same. There is sure to be sad loss among young ducklings, bred from debilitated stock. Good stock should be secured to start with, and when properly fed and cared for, there need be no fear of loss.

A good incubator carefully operated without variation of temperature should receive the eggs. They take twenty-eight days to hatch. Duck eggs will hatch well in any of the standard incubators; they require more airing than do the eggs of the hen, and I have found that by sprinkling them every other day, after the first week, I was sure of a good hatch. Sprinkle the eggs, or moisten them thoroughly, with warm water, when they are out of the machine, and do not put the water in the incubator. I found this much the best plan. I think wetting the shell of the egg helps to soften it and make it more brittle, enabling the duck to break its way out easily. I also do this when hatching duck eggs under hens.

A brooder adapted to chicks will answer equally well for ducks. The little fellows should be at least thirty-six hours old before taken from the incubator and placed in the brooder, which should be previously prepared for them by placing a board about ten inches wide a few inches from the front of the brooder forming a very small yard with a little water fountain so arranged that they can get their bills in but not their bodies. The birds should be confined to this small space in front of the brooder for the first day, or until they have learned the way into the hover. Bed the little fellows with hay, chaff or cut straw. Keep the pens clean both outside and in. The welfare of the ducklings depend upon this. Be sure to give them shade.

Mr. James Rankin has been called the father of the duck industry in America. He and a number of others in the East are now hatching by the thousands and tens of thousands. He writes: "With us it is the surest crop we can grow; it makes the best returns of any crop on the farm."

As he is a noted expert in the business I cannot do better than give his directions for raising the ducks and his formulas for feeding at the different ages. I have tried them myself and do not think they can be improved upon.

Feeding

The first food should consist of bread or cracker-crumbs slightly moistened and about 10 per cent of hard boiled eggs chopped fine, shell and all; mix in this food five per cent of coarse sand. Do not place grit by them and expect them to eat it, but mix the sand in their food and so compel them to eat it as it is the most essential part of the whole thing.

Scatter the food on a board, place the young ducklings on it and they will be busily eating it within ten minutes. One hundred to one hundred and fifty ducks can be put in one brooder six feet long. When two or three weeks old, not more than seventy-five should be kept in one brooder. The heat under the hover should be kept at about 90 degrees for the first day or two, when it should be gradually reduced as the ducks grow older. In the climate of Southern California, ducklings rarely require brooder heat more than two weeks.

The second day rolled oats and bran can be added to the food; a little finely cut clover, lettuce or cabbage can now be safely used. At ten days feed one-fourth corn meal, the rest wheat bran with a little rolled oats mixed in, not forgetting the grit, about ten per cent of ground beef scraps, and the same of green food. At six weeks Quaker oats, grit and ten per cent beef scraps; at eight weeks old feed equal parts of bran and corn meal with a little Quaker oats, grit and beef scraps, but no green food.

The birds should be ready for the market at ten weeks old. They should be fed four times a day until six weeks old, then three times is sufficient. They should be watered only when fed until six weeks old, then they should be watered between meals also. Feed at each meal all they will eat up clean, then take the remainder away; keep the pens dry and clean and be sure you give them shade.

For breeding birds, old and young, during the summer and fall, when they are not laying—feed three parts wheat bran, one part quaker oat feed, one part corn meal, five per cent beef scraps ground fine, and five per cent grit, and all the green feed they will eat in the shape of corn fodder cut fine, clover, or oat fodder, or alfalfa. Feed this mixture twice a day, all they will eat.

For laying birds—equal parts of wheat bran and corn meal, twenty per cent of quaker oat feed, ten per cent of boiled turnips or potatoes, fifteen per cent of clover rowen, alfalfa, green rye or refuse cabbage chopped fine and five per cent of grit. Feed twice a day all they will eat, with a lunch of corn and oats at noon; keep grit and crushed oyster shells before them all the time.

Mr. Rankin adds: "I wish to emphasize several points. Do not forget the grit, it is absolutely essential. Never feed more than a little bird will eat up clean. Keep them a little hungry. See that the pens and yards are sweet and clean, for though ducklings may stand more neglect than chicks, remember that they will not thrive in filth. If any one fails in the duck business, it must be through his own incompetency and neglect."

Mr. Rankin has his yards swept twice a week. These sweepings amount to many tons each season, and are spread evenly over his grass farm, giving enormous crops of good hay, so that where twenty years ago only six tons of hay were cut, now the crop is 125 tons.

Pekin Ducks

Something About Geese

Geese are, of all fowls, easiest to raise where grass is abundant, for they are grazing animals. Among the various breeds raised in this country the Toulouse is the most profitable goose to raise. It grows the largest, matures the quickest and is not so much of a rambler or flyer as the other varieties, and as it does not take so readily to water it grows more rapidly and accumluates flesh faster than other varieties, and is not so noisy.

There seems to be a steady demand for the beautiful large, gray Toulouse variety. They deserve every word of praise given them. They have been known to live to a great old age. I have had a friend in England who had a goose that had been for more than a hundred years in the same family, and even at that age produced as many and as fertile eggs as any in the flock. In fact, that goose had more broods each year than any other goose in the neighborhood.

There are many points about raising geese that can be learned only by experience and a little practice is worth a world of theory. Intelligent and systematic breeding is sure to bring both pleasure and profit to the breeder.

Hatching and Feeding

For hatching goose eggs if setting hens are used, keep them free from lice by dusting with insect powder every week, and put from four to six goose eggs under every hen. After eight days test-out, leaving four fertile eggs under every hen to hatch. Goose eggs should be sprinkled every fourth day after the twelfth. In damp or cold weather with warm water; 103 degrees in hot, dry weather, and float them in water from one and a half to two and a half minutes. If incubators are used, float always. At the last float hold the pip up so as not to drown the gosling inside the egg. If the gosling remains and dries in the shell, it should be helped out. Break away a little of the shell, and if the inside lining does not bleed the gosling is ready to come out. Ring out a cloth in water as hot as you can bear your hands in, wrap the egg in the cloth and leave for a few minutes. You will find the gosling will come out bright and clean. Keep the goslings warm until they are dry and can run around. When they are twenty-four hours old put them in a box, the bottom covered with sand, and feed them often with a crumbly mash of one-third corn meal, two-thirds bran and a pinch of sand.

Goslings Are Healthy

No other young in the whole tribe of domestic poultry is so up-to-date and healthy as a young gosling. Given a tender grass plot and a bit of warmth, it goes merrily on its way, nipping a living and asking favors of no one. They eat daintily, preferring grass to all other foods. With their chatter they are ready to meet you, take a few mouthfuls of foods, and, with the same old tune, they lazily saunter away in search of grass and more rest.

Geese are turned out to pasture just the same as cattle, their bills having serrated edges which enable them to graze. They never need a warm house. An open corrall is much better in California for them and they are not given to disease. Goslings, however, should be provided with shade, as they suffer from heat, getting a specie of blind-staggers or sunstroke if exposed to the sun.

One of the best items of profit to be derived from a flock of Toulouse geese is the feathers, which are clear gain, costing nothing but the trouble to pick them. Watch them in the fall and spring, twice a year, when they begin to pull out the feathers and thrown them away. I know then they are ready to pick. I think it is cruel to pick at any other time. Make cheesecloth sacks which will hold two pounds of feathers. Make them large, as the feathers will curl better if they are not packed together. Hang the sacks on a clothesline every sunny day for about two weeks, then keep them in a well aired room. Women living in the city will be your best customers providing you let them know you have good feathers for sale. One can get from 75 cents to $1.00 per pound, and can never supply the demand. The breeders should not be picked when they are laying.

The Varieties

There are a number of varieties of geese, but the most profitable are the Toulouse, the Embden, and the China. Of the latter there are the two kinds, the brown and the white. The color of the Toulouse is gray and white and the Embden is white. The Toulouse and the Embden are the larger. A pair of Toulouse have been known to weigh 59½ pounds, and an Embden pair has tipped the beam at 57 pounds. They are great layers of large eggs, of which they will lay thirty to forty a year, although I know a woman who has a goose that layed 70 eggs without wanting to sit.

In mating, allow two geese to one gander, though they generally pair off and the gander will stay with his actual mate nearly all the time. The gander is the protector of the goose, especially in breeding time. He will defend her and her nest fearlessly.

Hens as Mothers

It is not an easy matter to distinguish the sex. It is a good plan to put goose eggs under a hen. It takes thirty-one days to hatch them. Then you want to be on the watch. The hen will set all right, but when the young ones break the shell and the hen sees a queer, green little creature, with a long, wide bill saluting her, she takes it for a freak of nature, and off comes it head. Not many hens will claim the young geese or hover them; so take the goslings away as they hatch and try the hens, giving the goslings to a good slow, gentle hen. As soon as she takes them without any fuss there is no danger. If the weather is nice they should be turned out in a small enclosure, which can be changed every day or so. Use boards six feet long and twelve inches wide. After a week let them go, and their foster mother's trouble begins. The little goslings do not care for her calling; they are hustling for every spear of grass and she has to hunt them. Her business is to keep them warm at night and warm them in the daytime if they get chilled. Never allow goslings to get to water to swim until they are fully feathered, and then only let those go that you wish to keep for breeders. Many of them will do as well if they never go swimming. During this period you must keep the old geese away, as they will fight the hen and molest the young.

You cannot raise geese as you do chickens and ducks, on a city lot. They must have pasture. It is a wrong belief that geese or their droppings will kill grass or pasture. If you have a large flock of geese and a small pasture they will clean it up; that is, they will eat the grass as fast as it sprouts and give it no chance to grow, just as a cow on a city lot will soon have only bare ground and you will have to tie her in the road. If you do the same with geese you would find the grass growing again the same as before. Geese are easier to raise than any other young fowls.

Mrs. Sly's Well Arranged Yards

PART II.

1001

Questions

and

Answers

CAUSE AND CURE OF SICKNESS

Apoplexy—What is the trouble with my hens? They seem healthy and all at once they begin to gasp and fall over dead. I cut one open and it was in fine condition, fat and nice. I cannot make out what it is.—Mrs. C. S.

Answer—Your hen had apoplexy from being over-fat. The over-fat condition weakens the muscles, and the heart and brain give way. Give the whole flock a little Epsom salts in the water for a week, cut down the amount of grain, especially any corn or corn meal in their feed, and feed more green food and more animal food with, of course, charcoal and grit.

Air Puff—I have been a constant reader of your articles and find them very good but I have a case I never remember reading about; it is a Barred Rock about 6 or 7 weeks old. A few days ago it went to limping and I supposed it was some of the others crowding but I have since noticed its whole right side was puffed away out, just the skin, and I took a needle and made a small opening and there was nothing but wind in it. I repeated the same operation next day. It eats and drinks and aside from the limping, seems to feel all right. They have a nice clean run and lots of green stuff. I am feeding cracked corn, wheat and Kaffir corn. Could you suggest a remedy and tell me what the disease is?—Mrs. J.N.H.

Answer—Your chick had what is called "Air Puff" and you did just right in puncturing the skin; you saved its life by it. The trouble comes from a wound or abrasion of the lung tissue resulting from violence of some kind. After caponizing a chick this trouble often develops. I have seen the poor little things almost as round as a ball and so light from the air under the skin that the slightest breeze rolled them along. Chicks that get trampled on by their mothers, or cockerels that fight are liable to suffer from injuries that result in "air puff." They become inflated with air. The treatment is a good nourishing diet. I resort to bread and milk in such cases. It is easily digested, and, puncture the skin to let the air out. In slight cases where there is only a little air under the skin it will disappear gradually without treatment, but if there is a considerable amount of air it is necessary to prick the skin and let it out.

Bumblefoot—I have a lame hen; she limps on her left foot. She eats as well as my other hens, her comb is red and looks as healthy as the others.

I feed my chickens cooked vegetables in the morning and dry wheat at night. They have plenty of fresh water, and as I live in the country on a new place they have plenty of green grass and insects. I have kept chickens for three years and have never lost any of them only this way.

If you will tell me what is the trouble I will be very much obliged to you.—Mrs. M. M. C.

Answer—Your hen has probably what is called "bumble-foot." It is something like a stone bruise or a corn in human beings. It usually comes from a corn or bruises of the feet, wounds with thorns, broken glass, hard stones or other sharp substances. The ball of the foot becomes swollen, inflamed, hot and painful. The fowl appears in pain. Corns are often caused by too small or narrow perches, which compel the fowl to grasp them tightly in order to maintain their position. This firm grasp continued night after night affects the circulation of the part of the foot that comes in closest contact with the perch. A similar condition may be caused by heavy birds flying from their perches and alighting upon a stony surface or hard floor.

If it has not yet become an abscess, simply cut off the thickened skin or corn without causing bleeding and paint the corn with tincture of iodyne. If pus has developed, soak the foot in warm water twice a day and poultice until the inflammation is reduced. After thoroughly cleaning the foot, if pus has developed, open the abscess freely with a sharp knife and scrape out the diseased matter. Wash out the wound carefully with peroxide of hydrogen or carbolized water. Stuff the wound full of iodyne gauze and bandage it. Continue this treatment daily until the wound is almost healed, then apply a good ointment daily until it is entirely well. The bird must be kept on clean, dry straw until fully recovered.

Swollen Feet—Will you extend a helping hand to an old batch who is having endless trouble with a few chickens? They begin to get lame and after a few days cannot stand on their feet at all, and some of them have great swellings on top of their feet that look like a big boil. I only have about forty in all; they have all the range they want in abundance and wheat twice a day, together with scraps from the table. My hen house is log, 12x16 feet, plastered on both sides, two windows with glass 12-24. The roosts are about eighteen inches from the floor. If you can tell me the cause and cure I will thank you kindly as I feel sorely tempted sometimes to kill all of them and start over. They are just common hens.—D. W. M.

Answer—Your hens have either bumble-foot or rheumatism. The bumble-foot comes from an injury to the foot and is caused by hens jumping or flying down from a high place onto stony ground. It is also caused by rocky ground and is somewhat like a stone bruise or a corn in the human family. It usually occurs in heavy, elderly hens and your plan of killing them off for the table would be a good one. The cure is to lance the "boil" and gently squeeze the core out, then wash with peroxide of hydrogen and bind up with a soft rag and keep the hen on clean, soft straw, not allowing her any place to roost. Bumble-foot sometimes comes from sharp edges on the perch or very narrow perches. Discover what is hurting the feet and remove the cause. It is sometimes necessary to poultice the feet to draw out all the pus. Rheumatism usually comes from damp houses or damp ground and to cure that you have to change those conditions. You can also give the fowls a little Epsom salts in their drinking water, or give each affected hen one dose of Epsom salts (half a teaspoonful) in a little water and put into the drinking water half a teaspoonful of bi-carbonate of soda to a quart of water. But I think your plan of de-capitating them and starting with fresh young hens would be better than trying to cure them.

Bronchitis—Will you kindly tell me what ails my White Leghorn hen? She sits around most of the time and squacks and slings her head and when I hold my ear to her side I can hear a continual rattling. Her comb is red and she eats well. I feed corn, wheat, Kaffir corn and table scraps. They run on plenty of green range. Her nostrils are clean. Age 8 months.—C. C. S.

Answer—Your hen seems to have chronic bronchitis or is taking cold frequently. See that she does not sleep in a draught nor in a house that is too tightly closed. Give her a teaspoonful of honey night and morning for a week and keep her clean from lice and I think she will be well in a week. A little red pepper and chopped onions in her food would also help the cure.

Comb Discolored—I am a constant reader of the Tribune but for the first time I take the liberty to ask your advice. I have a White Leghorn cock two years old; he has always been healthy, but for the last two months I noticed that his comb and wattles turned a deep purple and would remain so for days, then they would change to a natural color again, but only for a day or so, and then turn purple again. He seems to be healthy and vigorous in every way. Now, can you tell me what can be the matter with him and what I can do for him, or if it would be wise to use him any further for breeding purposes?—Mrs. L. S.

Answer—The comb tells quite a little story of what is going on in the organs of the whole body. Any change in the appearance of the comb is indicative of a disturbance in some other part of the bird.

The dark colored comb is an indication of a disordered liver and indigestion. The dark comb is one of the first symptoms noticed in congestion of the liver and most cases of this come from an overfeeding of a ration too rich in starch elements, such as too much potatoes or bread in the table scraps, and insufficient exercise. I do not know how you are feeding your fowls, but I would recommend you to put a little Epsom salts into the drinking water, or you can give him alone a small half teaspoonful in a tablespoonful of water, and put in the drinking water of the whole flock ten drops of tincture of nux vomica to a pint of water. Feed plenty of green food and more meat than you are now giving; keep this up for a week and then turn the birds out on a grass range if possible, otherwise give the birds as scratching material the waste from an alfalfa hay mow and allow them only a little grain, wheat, and make them scratch hard for that. It would not be advisable to use the male bird for breeding. Breed only from the most vigorous stock you have.

Bald Headed—Some of my hens are becoming bald headed. The feathers for half an inch and more back of the comb disappear. The hens seem in the best of health and lay well. There are no lice or mites on the chickens, on the roosts or in the nests. If you can give me a remedy I shall consider it a great favor.—Mrs. E. E. C.

Answer—This is not at all an uncommon occurence just before the moult. Those feathers have merely ripened a little earlier than the others, and strange to say, it is usually the best layers that are so affected. You can grease the bald spot with a little vaseline. This will hasten the growth of the new feathers.

Blind Chicks—What is the matter with my little chickens? They are about two months old. I find them with one eye shut and sometimes both, and when I open it a watery substance comes from them. When only one eye is effected they are perfectly blind in it, but can see all right out of the other and when both eyes are affected they are blind in both.

Their mouths are perfectly clear and they have a rattle in their throat. They have been affected now for about two weeks and several have died. It seems very contagious.

I have put spirits of camphor in their drinking water and sulphate of iron. I also made a salve of lard and Egyptian insect powder and rubbed that on their eyes with a feather, which was very highly recommended to me, but everything has failed to cure them. They run on a yard of green grass all the time.—Mrs. A. L. S.

Answer—The starting point of nearly all cases of blindness in chicks is in roupy breeding stock. A slight chill or cold is sufficient to start an epidemic of this blindness in a flock of chicks, if they already possess the inherited tendency to weakness of these parts from parents that were not in fit breeding condition. This blindness is a result of an inflamation of the mucous membrane of the eye and lids which produces a sticky exudate which gums the eyelids together.

Sometimes the inflamation of the lids is excited by irritating substances like lime or sharp, dusty sand, insect powders or kerosene getting into the eyes. These causes may produce blindness in chicks that do not have roupy ancestors. That form of inflamation of the lids accompanied by hardening of the lids is not uncommonly caused by irritants, kerosene particularly.

Uncleanliness is another cause of blindness of this sort and too many who attempt to raise chicks are careless in this respect. Lice and mites also do their share to cause the trouble.

The best way to remedy such cases is to prevent them or remove the cause if possible. In cases where there is an amount of exudate it will be well to bathe the eyes with a solution of boracic acid, fifteen grains to a half cup of water, and then dry with a soft cloth and apply a little carbolic salve. It is difficult to get satisfactory results dosing young chickens with medicine, but you might give them either a little bread and milk with a sprinkling of red pepper and sulphur on it or rice boiled in milk with a tablespoonful of ground cinnamon for each pint of milk.

Mouths Red and Sore—I have about 60 little chicks, hatched out with hens, and am letting their mothers care for them. I feed wheat. Have a box of bran and feed meal before them all the time, fresh water daily, lawn clippings most of the time; a box of ground shells, and once a week ground meat or ground bone, fresh. Am I feeding them properly, and if not, please tell me how to feed? I let all the chickens run together, big and little.

For lice, I grease the old hens and

the little chicks' heads under the neck, so do not think the lice bother them much.

I have lost two or three little chicks about two or three weeks old. They eat, but get poor and weak, and finally die. Some of them keep swallowing all the time, and on examining their mouths, find them red and sore; white canker under the tongues. What can I do to keep the rest of the flock from getting it? Is it contagious? Will you please tell me what to do for them? Thanking you in advance, I remain, yours truly. —M. C.

Answer—Your little chicks have lice, to begin with. Lice stunts the growth and make them poor and weak and ready to take any disease that is going. Yours have taken canker. Now, the thing to do is to put four grains of sulpho-carbolate of zinc in an ounce of distilled water; paint the spots lightly with that solution one day. The next day swab or spray the mouth and throat with peroxide of hydrogen; using one spoonful of peroxide to one spoonful of water. Continue this until they are well. It is very contagious, and you had better put a little germazone into the water. Powder the chickens well with lice powder, and their mothers also.

Your feed is fairly good, although I prefer the chick feed composed of a variety of grains. This you can buy ready mixed at the poultry houses.

Canker—I am anxious to know if the heavy Black Orpingtons are hardy. I have just bought a fine cockerel and four hens; one of them has just got canker. What is the cause and remedy? They are kept in a yard by themselves and get clean drinking water, and sleep in a fresh air house with open side facing east. Do you favor open front houses for fancy breeds? I feed them with mash in the morning and wheat in the afternoon, and alfalfa grows in their yard.—Mrs. M. N.

Answer—The Black Orpingtons are very hardy. Am sorry your pen has canker. The cure for that is to paint the spots with sulpho-corbolate of zinc (four grains in an ounce of distilled water) night and morning. This will kill the germ, but in case it is

diphtheritic roup, would advise you to paint it one day with the sulpho-carbolate of zinc and the next day with peroxide of hydrogen, as the latter kills the diphtheritic germ. The open front houses are the best for every kind of fowl in this climate. A change of diet will often affect the droppings of the fowls, when they are normal. You had better slightly change the foods, or if you feed them charcoal, it will materially assist the digestion, and you need fear no trouble. A little Epsom salts in the water, if the fowls are very fat and heavy, is also an assistant, but by giving them plenty of green food, you will have no trouble.

Sore Throat—Will you please tell me what to do for our hens? They have something the matter with their throats. It gets sore right on the tongue. It is yellow and just like leather. It first starts like a canker; they cannot eat and droop around. We have not had any die yet, but I would like to know how to treat them.—Mrs. L. A. G.

Answer—It is either diphtheric roup or canker that your hens have I think it is the diphtheria, and the treatment is to swab their throats with peroxide of hydrogen. One spoonful of peroxide and one of water; or you might use the peroxide pure. I do not think it would be too strong. Give the hens also a one-grain pill of quinine as a tonic and put a little germazone in the drinking water.

Cold in the Head—Can you tell me what is the matter with my chickens? They eat, seem to feel good, sing and play and are laying good, but they seem to have a cold, or something. They try to blow their nose and bubbles come out. Have been that way for about six weeks; they have a good coop with no air holes; six by eight; one end open; only twenty-five to roost in it. They have had bluestone in their drinking water every day for a month; they do not get any worse or seem to be an better; they have warm mash for morning feed and wheat noon and night. Would they be good to eat in that condition? —F. C. H.

Answer—I am afraid that your chickens are too crowded in their roosting quarters and that they get too warm at night and come out into the cool morning air and in this way take cold. Or the open end may be towards the night breeze. They evidently have, for some cause, silght colds. Bluestone, or germazone in the water is an excellent cure and by adding chopped onions and a little red pepper to the mash, should cure them. One teaspoonful of red pepper for every twelve hens is the dose. If the chickens are not feverish and the discharge from the nostrils has no bad odor, would consider they are fit for food.

Cough and Sneeze—Will you please tell me what is the matter with my birds? I have several that cough or sneeze, I do not know which. They will shake their heads and "holler." One can hear them quite a distance. Will you please tell me the disease and remedy?—B. J., Tucson, Ariz.

Answer—Your fowls have bronchitis and perhaps some influenza. Give them bread and milk for supper, and a quinine pill and half a teaspoonful of red pepper mixed with butter. And see that they do not sleep in a draught or in a house where the rain comes in on them.

Congestion of the Brain—I kindly ask you to tell me what is the trouble with my White Wyandotte cock; he was strong and vigorous until he commenced to moult when he got bowel trouble, of which I cured him. One evening as I went to shut the chickens up I found him on the floor unable to get on the roosts; when I picked him up and set him on the perch he fell as though dizzy and rolled over and over. He continued to get worse until at last his head and neck turned completely around with the bill setting straight up. His appetite was good until his neck became so twisted that he could not swallow. As soon as he was a little disturbed he would twist his neck entirely around. If I put him up on a box no matter how low it was, he seemed to get dizzy and would tumble around on the ground. I feed mostly wheat, lucern leaves, bone and some meat fresh ground, mangle beets and grit. What do you think of whey for chickens?—Mrs. L. P.

Answer:—The symptoms you describe are of congestion of the brain. This

may occur in fat birds from fright or indigestion and is frequently associated with irritation of the intestines from worms. In case it is worms, give ten drops of turpentine in a teaspoonful of Epsom salts in a tablespoon of water. A pill of three or four grains of asafoedita will often cure this trouble, which really comes as a nervous affection. It also may come from bad meat. Give charcoal in the feed. Whey has not much nourishment for chickens, but you can give it them as a drink.

Catarrh—Can you please tell me what the trouble is when chickens cough and their nose runs, also state the best way to rid them of this plague?—Mrs. S. A. B.

Answer: Your chickens have taken cold and may probably have lice. Try to discover what is giving them their severe colds. It is probably some draught. Put a piece of blue stone in their drinking water (the size of a bean in a quart of water) and give them a pill of the following: Mix two tablespoons of lard, one each of mustard, red-pepper, vinegar; mix thoroughly, add sufficient flour and make a stiff dough. Give a bolus of this as big as the first joint of your little finger every night.

Cannibalism—I had a hatching of Black Minorcas three weeks ago of 115 chicks; today I have about 80. In the first place, the chicks are hearty and well, but will bite the rectum of the other chicks and in two or three minutes will just tear the bowels out and kill the little chicks. Every one will give it a nip and if we are not constantly on the alert all would be dead. No one of whom I have inquired has ever heard of such a thing. I have raised these just as I raise my White Leghorns. I hatched 160 seven weeks old and today have 158 fine chicks. You would oblige me very much with a remedy.—W. P. H.

Answer: The remedy for "cannibalism" is first, to keep all the chicks busy with exercising; in order to do this keep the floor of the brooder covered with chaff, or finely cut alfalfa hay at least an inch deep and feed the chicks small grains (chick feed) in this; the hay or chaff keeps the toes and feet covered, conceals them, and the busy little things are so occupied scratching that they do not get into mischief. Secondly, give them a little more animal food, or milk.

The cannibals have a craving for animal food, and sometimes a bit of fat salt pork, whether fed to them or nailed up where they can peck at it, satisfies this craving. Thirdly, find the first leader of this mischief, and either kill him or isolate him and give him to a hen to bring up. This bad habit is usually started by one chick, and all the others follow suit and soon the whole brooder will acquire the habit and it is almost impossible to stop it if it has got a good start.

Toe Eating—Can you tell me what causes little chicks to pick at each others toes? They all pick at one till the blood comes, then so many chase it that it dies. Then they start on another and sometimes they even eat the entrails out. I bought my chickens when they were a week old and fed them according to your directions. I first fed raw meat and cooked, then I tacked pieces on a board to keep them busy but nothing seemed to stop them and I took the one out with the sore toes. I gave lime and salts and charcoal. I hatched some dark colored chicks in my own incubator and with them I have not had any trouble in that way. I trust that you can help me. —H. L.

Answer: It is usually with the white or light colored chicks that we have this trouble. The little toes are so attractive and look so very good to eat that a lively chick will often try to taste his neighbor's toe and it tastes so good that he continues the performance and soon teaches the others. Dark toes are not so attractive looking, hence their immunity. You did quite right to add more meat and even a little salt pork to their diet, but the best way of preventing the trouble is to give the chicks chaff at least an inch deep in the nursery of their brooder. I have found that alfalfa hay or wheat hay cut in a clover cutter an inch in length make very good chaff for the chicks. I scatter the chickfeed a little at a time, three times a day in this and the chicks scratch in it and find the grains and at the same time it conceals their toes from their hungry brothers. In this way you not only prevent this vice but you make the chicks scratch many hours a day and that broadens their backs and develops the egg organs and strengthens their digestion, keeps them out of mischief, healthy, ha   and busy. Try this plan and you will be surprised to find what

extra fine layers you will have next year.

Cancer—The writer wished to know if poultry are subject to cancer.—J. H.

Answer: Poultry are not subject to cancer, but they are to tuberculosis, which may be taken for the same. There is no cure for this but the hatchet. A thorough disinfecting of the premises must be made. The bodies of any fowl dying from this disease should be burned, or buried very deeply, as it is an infectious disease.

Cough—We have a disease in our poultry. They have a phlegm in their throats and cough; they seem all right to look at them; they eat and drink until the day before they die, when they begin to droop. I notice it only when I let them out in the morning, or by disturbing them at night. They are fed about twelve pounds of wheat a day, two sheaves of barley, a pan of soaked bread, occasionally a feed of boiled potatoes mixed with bran and a little cayenne pepper. I have been giving them carbolic acid in their drinking water, about seven drops to a milk pan full; they usually drink it before being let out of the feed shed. We have lost only two birds, a peacock and a young turkey, but they all seem to have it. I will be much obliged if you can tell me what the disease is and how to treat it. —M. G.

Answer: Your chickens have a slight cold, more like bronchitis than roup. I would advise you to put some germozone into the water given them for drinking and some chopped onions in their food, and considerable red pepper. There is a possibility that their coughing comes from dust of some kind in their sleeping coop, or from barley beards in the straw. You had better not give them any more carbolic acid in the water. It is very injurious to turkeys. It is always best to try dieting and simple remedies. A teaspoonful of honey once or twice a day will often cure phlegm in the throat.

Crop-Bound—I have about 100 Leghorns; been very healthy all winter; laying good. Now, about six weeks ago I lost eleven of the heaviest ones in six days. They had yellow droppings; lived only two days and died. Four others died after having a heavy crop hanging down; they were apparently healthy and laying eggs regularly; I cut the crops off three of them and found nothing but long strings of hay. Please oblige me by telling me the cause and what remedies.—A. F. H.

Answer: Your hens are suffering from what is called crop-bound. They eat long pieces of hay, which form into a ball in the crop and cannot pass through them. After a time this ferments and decays and poisons the chickens, or brings on inflammation of the crop. When long pieces of grass or hay cause this trouble, as in your case, almost the only remedy is to cut open the crop of the bird and wash it out. Have someone hold the bird so you can have both hands free to work. Pluck enough feathers from the breast to give bare skin half an inch wide by two inches long. Then with a sharp knife cut through the skin, lengthwise of the bird, an opening one inch long over the place of the swollen crop. Cut only the skin, leaving the crop untouched until the blood of the first incision has ceased to flow. Then cut through the crop a little over a half inch long. Half an inch may seem short, but you will be surprised to see how large the opening is after you have worked through it for a while. In removing substances from the crop be careful to let as little as possible slip between the skin and crop; with a button-hook or anything else handy, remove the contents. If filled with grass or hay, it is sometimes necessary to cut the mass with scissors before any start can be made. When the crop is apparently empty, push your little finger into it, feeling to know whether there is any obstruction at the outlet. If you find the opening clear, the last thing is to sew up the cut. With needle and white silk thread, take two single stitches in the cut in the crop, then in the same way take three stitches in the skin, tying off the silk at each stitch. Be careful not to include the crop in the knot tied. After the operation feed soft food, omitting grain for a week.

Comb and Skin Turned Dark—We have White Leghorn hens and about four weeks ago one of them died; her comb and skin turned black, and water ran from the mouth before she died; she sat on the roost in the day time and was only sick one day. Today we had another hen to die with the same trouble; they do not seem to have any cold. We have a very comfortable hen house and large yard; now if you can tell me what the disease is and the cure for it, we will be more than pleased.—F. M.

Answer: The dark comb and skin and water running from the crop is an indication of liver trouble and also of poisoning . The poisoning may be from carbolic acid, lice killer, paint, phosphorus (rat poison), or ptomaine poison from bad meat. Unless I knew more of the symptoms, more of what care the fowls have, how they are fed, etc., it would be very difficult for me to recommend any remedy.

Yellow Blisters—The last week I have had three hens and a rooster break out on the exposed parts of the head, with what looked like yellow blisters, and in places it is as if it had bled and dried up. The first hen, a blooded Light Brahma of three years, I killed; the others I have penned up by themselves; have put carbolated vaseline on them. They seem healthy enough otherwise, and the breaking out seems to be only on the exposed parts of the head. Would you advise killing them, or can they be cured?—Mrs. C. M.

Answer: Your fowls have the chicken-pox. You have done exactly the right thing in putting carbolated vaseline on the spots. That will kill the germ of the chicken-pox, and in about a week they will be well. It is rarely fatal except with young chickens, and the cure is carbolic salve, applied about twice. A little sulphur in the food will hasten the cure.

Warts on Combs and Eyes—I am in trouble and I know you can advise me. September 24th, I hatched some Blue Andalusians. They have grown very fast, seemed extra healthy and vigorous until a few days ago, when warts began to appear on their combs and eyes. In one night they grew twice in size. I have nine and they are all becoming affected. What in the world is it and is it catching? They have run at large entirely and their feed in grain is mostly kaffir corn. They were such fine chicks and I was raising them for breeders, but now feel discouraged. I have a younger litter, four weeks old, but they are all right so far. My old birds are fine stock and very healthy. These warts did not make their appearance until the chicks were eight weeks old.—Mrs. H. E. S.

Answer: Your chicks have the chicken-pox and, I fear, of a virulent type, as the warts are growing so quickly. The cure is, to anoint the warts with carbolic vaseline or carbolic salve. Give them nourishing and easily digested food, adding a little sulphur, about a teaspoonful per day, and some chopped onions to their feed. It is very infectious, but rarely fatal except in the case of young chickens. Keep close watch on the smaller brood and apply the carbolic salve the moment you see a spot on the head. After a few days if weather is warm, you can wash the spots with warm soapsuds and a few drops of carbolic acid. This will hasten the cure.

Old Hen Dumpy—My old hen, the mother of all my flock, is ailing. She has not layed for nearly four weeks although her comb is quite red. She is a Barred Plymouth Rock. She is dumpy, does not care to eat or drink; droppings green and cream color, very loose. I have given no medicine of any kind: feed a dry mash mixed according to your formula, also rolled barley, wheat, table scraps, plenty of greens, oyster shell and they have charcoal, but do not seem to want it. Have kept chickens two years; they have always layed well; have had no sickness until now and they have no lice or mites. I turned the hen loose about two weeks ago; at first she seemed better, but now gets weaker every day. I boiled rice and milk, added cinnamon as you suggested, but she would have none of it. I do not know how to make a hen eat what she does not want. I do not know what to give her and would not know how to make her take it if I did.—M. K. L.

Answer: I think you had better give your old hen ten drops of laudanum in a teaspoonful of castor oil, and repeat the dose if necessary, in two days. Cut some bread into the size of dice and soak it in milk, sprinkling it with red pepper, then take the hen on your lap, holding her head with your left hand and with your right hand put one of the dice into her beak and holding her head up, gently push the soaked bread down her throat. Add finely chopped onion to this and turn her out on the grass away from the others. The onion is a good liver tonic.

Feather Pulling—Will you kindly tell me the cause of chickens pulling feathers from each other and eating them? We feed them wheat, cracked corn, etc., also ground bone.—G. H. T.

Answer: Various causes have been assigned for this habit, the most probable being improper rations and idleness. In some instances it is caused by mites or lice. As in some cases, the habit is due to insufficient animal matter in the rations, or to feeding too long on a single kind of grain, particularly corn, one of the first measures adopted should be a well balanced ration, containing skim milk, meat bone, vegetables or green feed and frequently varied. The Geneva, New York, experiment station applied to the feathers lard or vaseline in which powdered aloes had been mixed. After continuing this treatment for some time the habit disappeared, due to the disagreeable taste of the aloes. The skin and feathers should be carefully examined for lice and mites and if these are found the remedies recommended for such parasite should be applied.

Heart Trouble—I have a very fine rooster two years old. For the past two months he has been troubled by some difficulty in breathing. At times his comb and wattles become purple for two or three minutes, then the color gets red again. I have looked for canker but cannot find anything that seems wrong. Have used vaseline but it has not done any good. It seems to me more like asthma or bronchitis. Wish I could cure him for he is a valuable bird.—Mrs. I. G.

Answer: I am sorry to say that your bird has heart-trouble. This has been brought on by some great excitement such as fighting, fright or being chased. It may possibly be fat on the heart, which weakens that useful organ. You might try giving him in the drinking water nux vomica and sulphur comp. 2x twelve tablets to each pint of drinking water. Be careful to give him plenty of green food and grit, besides his ordinary food. Cases of this kind are almost incurable, but the treatment I have indicated may help him and prolong his life.

Indigestion—What is the matter with my late hatched chickens? They seemed all right till seven or eight weeks or age when the craw seemed to fill with water and they refuse to eat except when given meat scraps chopped fine. By holding the head down and pressing on the craw, a frothy substance will run from the mouth and the craw seems to be empty but it will fill again soon as before and they die apparently of starvation.—Mrs. P. T.

Answer: Your late hatched chicks are dying of indigestion. It comes from something wrong in their feeding or care. The usual causes are a lack of grit and not enough green food, but there are a number of other things that give it; such as too free a use of liquid lice killer or of distillate oil in their coops, sour or fermenting food, bad meat scraps, taking cold and lice which later weaken the chick so it cannot digest. Not knowing how you feed or care for your chicks I cannot say which of all these is the cause of your trouble. Give the chicks granulated charcoal and remove the cause whichever it may be. The meat was the best thing you could give them under the circumstances, as it is the most easily digested .

Indigestion and Liver Complaint— My hens are on a strike, and their faces and combs are becoming pale or yellow. What is it?—T. S. B.

Answer: You have been over-feeding, and now your fowls have indigestion. Indigestion in fowls is the cause of many ailments. With your birds it has been brought on by lack of grit, with not sufficient roughness (or filling) and too little exercise. How can indigestion be prevented? By dieting. Feed more bulky foods, such as alfalfa, and less solids. A continued grain diet of wheat, corn, barley, if few in quantities and not varied by bulky foods, vegetables, etc., will bring on indigestion, especially when but little exercise is taken. An insufficiency of clean water is also conducive to this trouble. Clover, alfalfa, any of the green stuffs or vegetables, usually fed to fowls, are absolutely necessary preservatives of health. Now, as to a remedy: Your fowls' indigestion has taken the phase of biliousness. Give each affected hen one of Carter's Little Liver pills, and give the whole flock a teaspoonful of baking soda in a quart of water every day for a week. Give no other water. Why do I recommend soda? Because, it helps to emulsify the too much fat in the bowels. You might give a teaspoonful of Epsom salts in the water

for a week, to carry off the bile which is overflowing into the intestines and being taken into the system. It is not kindness to feed your fowls every time they come near you. It is far kinder to keep them working for it and so keep them healthy.

Inflammation of the Crop—I have a Buff Orpington hen that has a disease I have never seen before. Her craw is swollen to several times its normal size and is filled with wind or gas. She eats but not as much as she should and is getting thinner all the time.—H. Y.

Answer: Your hen is suffering from inflammation of the crop. This is like a very severe attack of indigestion. The causes of this are irregular feeding or too much food being taken at one time. Partially decomposed meat, or putrid food of any kind will also cause congestion and fermentation of the contents of the crop. The same disease occurs when birds eat substances containing phosphorus or arsenic, or rat poison. The feeding of too large a quantity of pepper or stimulating "egg food" in the mash will also cause inflamed crop as well as trouble with the egg function.

Treatment—A clean, dry pen should be provided for the affected bird. Empty the crop of its irritating and decomposing contents by careful pressure and manipulation while the bird is held with its head downward. When the crop is freed of its contents, give two grains of subnitrate of bismuth and one-half grain of bicarbonate of soda in a teaspoon of water. The bird should then be kept without food for eighteen hours and then fed sparingly upon easily digested food, such as bread and milk. Half a grain of quinine morning and night for two or three days will complete the cure.

· **Necks Squirm—**I have a pen of beautiful White Rocks. I find three of them go off by themselves and mope. They drink a lot of water and when they do not eat their necks squirm about like a serpent and their crops look full but there appears to be nothing in them. I feed mash in the morning, warm with a little salt; wheat at noon, cracked corn at night; plenty of shell and charcoal; lettuce when I can get it and table scraps. I have raised them by hand and after all my trouble I do hate to think that I may lose any. Thanking you in advance.—Mrs. F. D. D.

Answer—The symptoms you describe are those of indigestion and may be caused by intestinal parasites, wrong feeding and lack of exercise. I would advise you to give them more green food, also mix a teaspoonful of terpentine in a quart of water and use that for mixing with the mash; also mix turpentine in the same proportion in their drinking water; keep this up for a week. It will kill the worms. As a tonic give them 10 drops of nox vomica in a quart of drinking water. This will improve the digestion, and strengthen them.

Influenza—I am in trouble with my chickens. Five of them have died since Monday. They open their mouths and gasp for breath and sneeze and their eyes are very watery. I feed wheat, cracked corn, plenty of green stuff and table scraps and they have a good run. I always wash out their drinking pans and rake out under their roosts at least every other morning.—Mrs. J. F. S.

Answer—Your chickens have influenza. They are taking cold in some way. Either there is a draught in their house or the rain comes in on them; a few have had the cold and they are giving it to the rest. Keep blue-stone in their water, and give each of them a bolus of the following, night and morning: Mix two tablespoons of lard, one tablespoon each of cayenne pepper, mustard, vinegar; mix thoroughly, add enough flour to make stiff dough; roll out; give a bolus as large as the end of your little finger. Put carbolated vaseline up their nostrils and in the cleft of the mouth, and give them chopped onions in their food.

Green Droppings—I have a White Rock pullet eight months old. She is dumpy, does not care to eat, her droppings are grass green and cream color and very loose. I feed alfalfa, cabbage, lettuce, beef-scraps, blood-meal, bone meal, wheat, kaffir corn, cracked corn and they have plenty of sand. Sometimes I put salts, soda and bluestone in their drinking water, and sulphur and red pepper in their mash.—Mrs. D. A. S.

Answer—I think you are giving your pullet too much medicine, and have upset her digestion. Put her by herself, give her rice boiled in milk with a little cinnamon added and sharp grit and charcoal. Sand is not coarse enough for hens. Also give her green crisp lettuce. Green food does not give hens looseness of the bowels but keeps them in good health.

Liver Trouble or Poison—I want your advise and a remedy for my sick fowls. The symptoms are, briefly stated: Grown chickens affected droop for two days, comb turns black and they die. Have lost nine in two days.

My chickens have free range, fresh water and plenty of barnyard scratching with Egyptian corn every night.—C. V. N.

Answer—The symptoms you describe denote either liver trouble or poison. In your case I think perhaps it is poison, either from rat poison, gopher or some poisonous weed. You had better hold a post mortem examination on the next one that dies and then you will be able to tell just what the trouble is.

Poisoned—Yesterday morning I found nine big chickens in my yard dead and about twelve more are dying. What is the cause? They sit on the ground, do not eat and the head hangs loose on the ground. The comb is dark and in the throat is a sticky slime like white mucilage. No bad smell; sometimes they jump a foot and lay down again. I fear they will all die. To a few I gave a teaspoonful of olive oil, and to some others fresh milk. I cannot imagine what it is.

Other fowls in the next yard are not affected, and all had the same food.—Mrs. F. C. P.

Answer—Your chickens have limber necks from ptomaine poisoning. Give the whole flock hypo-sulphite of soda; dissolve one teaspoonful in a quart of drinking water. And to each chicken that is affected give a piece of asafoetida about the size of a green pea. Use the gum form, and repeat the dose the second day. This disease usually comes from severe attacks of indigestion, caused by eating bad animal food, or the decaying carcass of a dead animal. Putrid meat or putrid milk will cause it.

Limber Neck—We have between 200 and 300 chicks two months old that are badly afflicted with limber neck, and we cannot find out the cause. The first two or three weeks we fed them millet and Johnnie cake made stiff and dry, of course corn meal, but they began to get sick, so changed to dry food, consisting of cracked wheat, millet, beef-scraps and grit, but the chicks still got no better, so now we are using just wheat and grit. They have lettuce every day and often young vegetables—tops and all. Until about a week ago they were kept by themselves in wire pens, but as an experiment my husband let them out to run and still they get sick. They do not all die as I bring them to the house as soon as we find the sick ones, but from one to seven die nearly every day. They have fresh water every morning. I do not try to doctor them, but just keep them warm. I have saved some pretty sick ones in that way. They are such a bother and we have lost so many in that way. The flock which is the most affected had a habit of huddling when they were small, until they would sweat and sometimes die. Do you suppose that could have anything to do with the present troubles?—Mrs. F. L.

Answer—Limber neck is due to a disorder of the nervous system and is usually the result of disturbances of the digestive organs from severe attacks of indigestion or from infestation with worm parasites. Chicks are sometimes affected in this manner by unusually hot days and nights. I think very probably their digestive organs were weakened by being overheated when they huddled and I would give the whole flock plenty of charcoal to eat, with plenty of green food and animal food, and no millet, as millet is very hard to digest. Give the sick birds a small piece of gum asofoetida, about the size of a green pea. Repeat the dose the second day. This will usually cure. Feed them with bruised garlic or with chopped up onions. Give them grit or very coarse sand in boxes to assist in the digestion, and I think you will have no further trouble.

It is possible that your chickens have worms. You had better open

the next one that dies and examine it and if you find it infected, give the others turpentine in the drinking water, half a teaspoonful to a pint of water (giving no other dringing water) or if you prefer it give a teaspoonful of Castor oil with ten drops of turpentine in it to each sick chick. The chickens dislike the turpentine in the water but it will kill the common round worms if continued for a week.

Sits Down to Eat—I wish to consult you about a hen. Her appetite is good, her comb is bright and healthy looking, and she gets about as spry as her mates, but she invariably sits down to eat, whether soft feed or corn. I have shut her off from the others and given her a condition powder in one meal a day; she has begun laying within a week or two but continues weak in the legs.—M. L. B.

Answer—Your hen has what is called "leg weakness,' and is possibly suffering also from greediness. If you had mentioned the breed I could have told more easily. If she is of the American breed, possibly she may be over-fat; if Mediterranean, she may have a slight amount of rheumatism. In any case, give her a little bi-carbonate of soda in her drinking water; a teaspoonful to a quart of water; also some chopped onions in her food and make her scratch for every grain she eats.

Leg Weakness—I am in trouble over my White Rock chickens. I only have a few, so would like to save them. When they are about three weeks old they get weak in the legs, and after a week or so they begin to tremble like a person that is nervous. They eat well until the last. I feed boiled egg and bread crumbs. They have green barley to run on. I feed kaffir corn at night. During the day I feed onions and table scraps. If you could tell me what to do I would be a thousand times obliged.—Mrs. W. K.

Answer—Your chickens are suffering from what is called "leg weakness." Leg weakness comes chiefly from wrong feeding, also from overcrowding at night and overheating.

Young chickens should either be allowed free range with a hen or be encouraged to work and scratch for their food. This strengthens their legs. The green food should form at least one-third of their diet and for such young chickens it would have to be chopped up finely. They cannot peck off sufficient green barley. It soon becomes too tough for them. The cure for leg weakness is a little tonic (a few drops of iron in their drinking water) and plenty of green food and cracked wheat instead of kaffir corn. If it comes from over-crowding or overheating, either under a hen or in a brooder, you must rectify this. See that they have "chick grit and charcoal."

Mange—I have a Plymouth Rock hen that has the under part of her body and legs and feet covered with hard, scaley sores of all sizes from a bean to a couple of inches across. Some are light yellow, some red and some purple in color. She seems to be all right otherwise, eats good and comb and head look red and healthy. Please tell me what ails my hen and if I can cure her.—Mrs. A. H. S.

Answer—I think your hen has mange. I would advise you to kill her and bury deeply or burn the body because when it is as virulent as you describe, it would be very difficult to cure and all those kind of diseases are exceedingly infectious. Carbolic salve at the first might have cured her but now it is too late and the time, trouble and expense of treatment, with the probability of the others becoming affected, would not pay.

Nervous Trouble—My cockerel, which I want to mate with my hens, has some sort of a nervous trouble that I have not been able to overcome. When he gets excited upon seeing a stranger or any sudden fright he jerks his head to one side as if he had St. Vitus dance. Will he do to breed from? Also I have two hens that shake their heads as though there was something wrong with their throats; they eat and sing and lay and their heads are red but I do not seem to cure them. When I swab with peroxide of hydrogen, they get better, but their breath seems to be short when they get excited. They are both a little too fat. I feed wheat, oats, corn, beets, carrots and a dry mash composed of two parts bran, one corn meal and one of beef scraps; have an alfalfa lot to turn them in on and have

never failed to get a heavy egg yield.—Mrs. M. A. G.

Answer: I would advise you not to mate a cockerel with any nervous trouble to your hens. This trouble is usually caused from indigestion, parasitic worms, defective propagating organs or weak heart. I do not know of any certain cure, but you might try giving him some asafoetida—a piece of about 4 grains every night for a week. I would advise you to eat him instead of breeding from him. Your two hens are evidently too fat. The shortness of breath denotes weakness of heart from over-fat. You are too generous with your corn.

Naked Chicks—Thinking perhaps you can help us I will ask you for a little of your time. Late in October we bought a hen caring for thirty chicks. We have fed them cracked corn, meat scraps, plenty of green stuff, charcoal and grit. They feathered out but since many of them have become bald, and the feathers fall from their neck and they are growing thin, still their wing feathers are long, making them look very queer. They are not incubator chicks, and we have examined them closely for mites, have dusted them for lice and they are quite free from either. What do you think is the cause and what can we do for them?—H. A. S.

Answer: Your chickens are huddling at night, crowding too closely together. This makes them sweat and their feathers fall out. Put a little carbolated vaseline on their heads and cut the feathers of their wings as close as you can without making them bleed. Give them wheat and more meat in their food and try to prevent their crowding at night. It is the crowding and lack of wheat in the food, lack of protein, that prevents the feathers growing, and the sweating makes them fall out and will make the chickens thin.

Over-Heated—I have a lot of young Plymouth Rocks. Some of them are as bare as my hand and a few have some feathers on them. They were hatched in an incubator. Can you tell me if they will ever be any good, or what do you think is the matter with them? They are not growing like my other ones. I feed them wheat, cracked corn, rolled barley, scalded alfalfa, beef scraps and sometimes a mash made with bran, bone meal and shorts.—Mrs. E. M.

Answer: Your chickens, when quite young, have been over-heated in some way, either in the brooder, under a hen or in the sun, and this has arrested the growth of the feathers and stunted them. They will never amount to much, and by greasing them with vaseline and continuing to feed them with the good food you now give them, they will, in a short time, feather out, but will not be as large as the other ones whose growth has not been checked. I advise you to fatten, kill and eat them as soon as possible.

Enlarged Liver—I have noticed a hen moping and eating but little for two or three weeks, but as I had broken some up from sitting, thought it the result from broodiness. However, as she got no better I separated her from the others, but yesterday she died. This morning I did as you advised, and duly performed the autopsy. I saw at once on making an incision what was the matter. Her liver was so enlarged that it occupied almost the whole cavity. I never saw one such a size. It was covered in blotches of pink spots, small as a pin point. There was fat around the heart and gizzard and layers of fat around the intestines; perhaps a fifth of an inch thick. There was plenty of grit in the gizzard but no food. The heart seemed in good condition, the body a good color, and flesh firm. In the cavities of the back is a substance, of which I do not know the name, that seems to be enlarging and hardened. There were many eggs but very small and undeveloped. Is this the kind of liver which is used as a delicacy and produced by over-feeding? My fowls were fed corn all winter and were much too fat this Spring. In March they had layers of fat an inch in thickness. I did not suppose that a laying hen ought to have any fat inside of her. How should that be? I have been reading over with great care your article on "Assisting the Moult." I have about 40 hens evenly divided between Plymouth Rock and White Leghorns. They are dropping off in their laying and beginning to lose feathers. Shall I begin now to reduce the feed, or wait until the middle of August? Should the roosters be put on the same diet as the hens? I have three acres in vineyard and corn, and think of turning the whole flock out in about a month. Will it be advisable

to let the roosters run with them, or should I keep them separated? I have a very nice, young White Minorca cockerel. I think of mating him with ten White Leghorn hens; meanwhile had I better let him run with the pullets, or shut him off alone? Do you think when feeding full rations, it is too much to give the noon mash? Would you think two feeds a day preferable? Do you advise giving the mash morning or evening?—G. S. H.

Answer: Your hens certainly had fatty degeneration of the liver, or the disease which the over-fat geese have when their liver is considered a delicacy. She simply had been fed an unbalanced ration containing too much of the fat element, and being a Plymouth Rock, had become over-fat. The substance in the cavities of the back is the kidneys. There are three lobes of these on each side. Your fattening ration had also affected them. So much fat will also affect the egg laying, will make small eggs and chickens will be weakly, as there will be preponderence of fat in the eggs from which they are hatched. A laying hen should not be anything like as fat as those you describe. About assisting the moult: I think as your hens are dropping off in their laying, you should commence immediately with their period of fasting. With hens in the fat condition in which yours appear to be, the fast will ultimately prove very beneficial. The roosters should be put on the same diet as the hens. It will be advisable to separate them, leaving only two hens or pullets with each made bird. The young white cockerel should be kept entirely away from the pullets and hens until you get ready to mate him. Shut him off where he cannot see any of the females. I think it is too much to feed three times a day, giving a mash at noon. Two feeds would be preferable. I prefer giving the mash in the evening, because then the hens go to roost with their crops full and in the morning are ready to commence scratching and working, which will induce egg laying.

Tumor—I have a few Leghorns, also Barred Rocks, and keep them in separate pens. One of my Leghorns is badly swollen in her lower parts; that is, where the egg bag is. Her bowels are loose and what passes from her is rather thin and whitish in color. Eats but very little, but her comb is nice and red; walks with great difficulty on account of being so very badly swollen. Am a great lover of chickens, and it hurts me to see her thus and not know what to do for her. If you will offer me a suggestion I will deem it a great favor.—C. J. W.

Answer: Your hen has either a tumor or a rupture of the oviduct, and there is no cure for it. The better plan is to kill the bird and save her suffering.

———

Tumor and Dropsy—I had a White Leghorn hen die a week ago from an ailment which puzzles me. Have looked through what poultry books I have, but can find nothing touching it. The hen was swollen between the legs to an unusual size and got so bad it could not walk. Finally it died and, upon opening it, at least a quart of water came away. The intestines were joined together in one solid piece. Can you tell me the cause and cure, as I have a Hamburg hen developing the same symptoms, and would like to save it if possible?—J. L. W.

Answer: Your hen died of dropsy, combined with a tumor, probably ovarian. There is no known cure for this, as by the time it becomes visible, the disease has progressed too far, and is usually only discovered after death. Some hens seem more subject to this complaint than others, and I would advise you to get in fresh blood and keep the hens healthy by feeding an abundance of green food. The cause is obscure.

———

Ovarian Tumor—I had a nice Orpington hen; she had been laying each day and appeared to be perfectly healthy; comb red, went around seeming quite well. I feed cracked corn and wheat, table scraps, and the chickens have goon range and plenty of good food. About four days ago the Orpington appeared to be lame in the right leg. I caught her, examined the foot and leg, could see nothing wrong and she continued lame, and with difficulty got on the nest. To all appearances the leg was broken, as it was harder for her to walk each day. Rather than see her suffer I had her killed. I dissected her; she was very fat with an abundance of eggs, one soft shell. I found in the right side of the back a growth about the size of a pigeon egg, which appeared to be part of the egg bag. The liver and other organs appeared to be

healthy. I hope that you may be able to tell me what the growth was and if there is a cure for it, in case any of the other hens have such symptoms. The hen was about two and a half years old. Would age have a tendency to hinder her?—Mrs. H. R. B.

Answer: Your hen had what is called an ovarian tumor. The trouble is very common, and yet we don't know very much about it. I am inclined to think that if investigations covering a large number of fowls kept under a variety of conditions were made, it would be found that cases of tumor like this are more abundant among fowls kept closely confined, or fed heavily for egg production, than among those kept under more natural conditions. It is quite reasonable also to suppose that the offspring of hens heavily forced for egg production would show weakness of the reproductive system, resulting in diseases of this character. It possibly also may come from an injury of some kind. Undoubtedly some strains or families are more subject to it than others. There is no cure for it and the only preventive is to keep the hens healthy and busy.

———

Tribulation—I wish a little information in regard to a Leghorn hen that died yesterday. She apparently choked to death; made a queer noise. We opened her and found at the bottom of her egg bag a large clot of black blood. Can you tell me what it was and if there is any cure for it? I also bought 50 young Leghorn chicks; they were lousy and most all had stuck-up behinds. I feed them on chick food and boil the water; I have about 18 left; this morning one began to twist his neck around and was droopy; I gave him Eosom salts. Please tell me what to do.—E. R.

Answer: Your White Leghorn hen had a hemorrhage of the oviduct; this is excited by any of the causes which lead to congestion and inflammation and may be contracted by green feed and the suppression of egg foods, stimulants, red pepper, etc. It sometimes occurs from trying to pass too large an egg. There is no cure than I know of, as death occurs before one finds out what is the matter. About the little chicks: They were badly hatched and I am afraid you will lose them all. Incubator hatched chicks having lice shows a filthy condition of the incubator and neglectfulness on the part of the hatcher. Give the chickens that are left rice, boiled in milk, adding a tablespoon of ground cinnamon to each pint of milk. Give them also an onion chopped fine and plenty of green lettuce. Give this in addition to the chick feed.

Paralyzed—One of our young chickens took sick on Saturday and by Sunday afternoon its legs and wings seemed to be paralyzed. They were stiff and it could hardly walk, and when it stood still it just shook as if it was cold, and it died Monday morning. We thought it may have been poisoned in some way. We have been feeding our chickens young cabbage leaves. Do you think that hurts them? We had two chickens die in the same manner and I think two more are going the same way. They stand around about two days and don't care about eating; and about the third day are awful sick, and the fourth day they die. Maybe you can tell me what was the cause of the two dying so we can save the rest.—Mrs. J. E.

Answer: I wish you had held a post mortem examination of the first chicken that died. It might have saved the lives of the others. With your meager description of the symptoms, you give neither age nor breed, nor tell the way they are fed, or cared for. I can but hazard a guess at the cause. I think it is worms in the intestines. The remedy is for the whole flock. Thirty drops of turpentine in a pint of water, allowing no other water to drink. Give this fresh as often as necessary for a week and it will usually exterminate the worms and cure the chickens. It is either worms or some kind of poison, but not knowing how you feed I cannot tell you whether it is poison or not. The young cabbage is very good for chickens unless you have been poisoning to kill the cabbage worms.

Abnormal Growth—I have a hen, and its crop hangs down so far that when it walks its feet are always hitting it. We cut it open once and only the corn and feed it had eaten came out of it. I have thought I would kill it, but I was afraid it might be a tumor and that the hen would not be fit to eat. She seems healthy otherwise. I want to ask you if turkeys get off the nest as often as the hens do when they are setting, for my turkeys never get off the nest unless I take them off. What is the best to feed young turkeys?—Mrs. J. A. M.

Answer: Your hen has a pendulous crop. This is usually caused by overfeeding of mash at some time in her life. It sometimes can be cured by a surgical operation. I would advise you to kill and eat the hen, as in time the crop will become sore. You can easily see before you eat it if a tumor has developed, in which case bury it. Turkey hens are very faithful setters. Some will starve to death on the nest so it is best to take them off every day. You will find full instruction for feeding young turkeys in an article in this book.

Diphtheric Roup—Having derived many useful ideas from your writings I take the liberty to ask your advice regarding a disease which has come upon my chickens. The first symptoms seem to be a sneezing or squawking sound as if the chicken had a beard in its throat; then a white membrane forms over the windpipe and the eyes close up and lumps break out around the comb. The lumps finally break and the eyes and nose run. Both Barred Rocks and White Leghorns are afflicted. The Barred seem to suffer the most.—Mrs. R. F.

Answer: I am sorry to say your fowls have diptheric roup. It is a very infectious disease and if you have childred you had better keep them away from the fowls. Spray the mouth, throat, nostrils and cleft in the mouth twice a day with peroxide of hydrogen. Give the fowls a quinine pill, four nights in succession, and once a day a bolus of the following mixture: Two spoons of lard, one each of mustard, cayenne pepper and vinegar; mix thoroughly, add flour enough to make stiff dough; give a bolus as large as the first joint of your little finger once every twenty-four hours. Put a piece in a quart of water, and allow them no other drinking water for a week.

Rhuematism in the Feet.—I have a very fine Buff Leghorn rooster and he seems to have rheumatism in his feet. Do you know any cure?—Mrs. J. M. S.

Answer—Rheumatism may result from long exposure to cold and moisture; it may be produced by overfeeding of meat; induced through the under-feeding of vegetable food and is helped along by previous rheumatic tendencies of ancestors.

Treatment—Bathe the feet and shanks with the following: One cupful of vinegar, one of turpentine and a heaping teaspoonful of saltpeter, mix in a bottle and shake well before using. For internal treatment there is no better remedy than Iodide of potassium. This is given in the drinking water, fifteen grains of Iodide of potassium to every quart of water. Give in small dishes so that it all may be used while fresh and thus avoid waste from having to throw away any, because it is mixed with dirt. Common cooking soda, one level teaspoon to each quart of water, or salicylic acid, one grain a day, has given good results, but the Iodide is the best and most satisfactory. Give plenty of green food.

Roupy Catarrh—What must I do for my young chickens? They are White Rocks, three months old, and I think a great deal of them. About one week ago I noticed one of them act as if it was blind in one eye and on taking it up to see the cause I noticed that one of its eyes seemed sunken in the head and badly swoolen all around that part of the head. A friend told me it was sore-head, and to kill it at once, which I did, but now find two others of the flock affected the same way.

I write to ask you to kindly help me and tell me if I should kill them, or if you know of a remedy for the disease.—A Reader.

Answer—Your chicks have "swell head," or roupy catarrh. It is an infectious cold, and they should be separated from the rest and treated immediately. Mix on tablespoonful castor oil; half teaspoonful each of turpentine, kerosene , camphorated oil and four drops of carbolic acid. Squirt a drop up each nostril and into the cleft of the mouth and rub the whole head with it. Also give the sick chicks a pill of quinine and one of asafoetida (one grain of each) and half a teaspoonful of cayenne pepper. Repeat every other night.

If the shoulders or under the wings are soiled with the discharge from the eyes, wash off carefully with a disinfectant.

Rheumatism—I have some chickens who are down in the legs. It looks like rheumatism. They can hardly walk and when they eat they sit down with the leg extended out in front from the knee; they cannot sit down like a chicken on a roost. What is the trouble and what medicine shall I use?—W. B. B.

Answer—Your birds have rheumatism. It is usualy caused by exposure to cold and dampness. The buildings where the fowls live and roost should be thoroughly dry, free from draughts and well ventilated. Affected birds should have frequent change of ration with plenty of green feed. Begin treatment with a dose of Epsom salts, 20 to 30 grains. The following day add a teaspoonful of bicarbonate of soda to the quart of drinking water. Bathe the affected joints with the following linament: Mix half cup of turpentine with half cup of vinegar and add a heaping teaspoonful of saltpetre; shake before using. The birds which are seriously affected would be better killed than treated.

Swell Head—I have decided to ask your advice about my Buff Orpington chickens. Their eyes are sore and swell completely shut. One hen's eye seems to be gone.—Mrs. E. E.

Answer—Your hens have what is called "swell-head" or sore-head. It is a species of roup. They are evidently taking cold from sleeping in a draught or from being in too hot a house at night and coming out into the cool morning air whilst still warm, or from lice which carry the infection from one to the other. The remedy is to first cure the evil, that is stop the draught or get rid of the lice, then bathe the head with per-oxide of hydrogen and water in the morning and rub it over with carbolic salve at night and give a pill of quinine (one grain) every night for a week.

Feed plenty of succulent green food and meat.

Specks of Blood—There is a speck of blood in the whites of the eggs laid by some of my chickens. Would you please tell me what causes it and what can be done for it? I have about two dozen hens in the back yard, and with the exception of a few table scraps, they are fed on wheat and a good balanced ration. They seem healthy, have plenty of green feed, but are, I think, fatter than they should be.—Mrs. E. P. G.

Answer—Blood spots in the albumen of the eggs results from a slight hemorrhage which has generally occurred in the upper two-thirds of the aviduct. They are the result of great functional activity and congestion of the blood vessels of the reproduction organs. They are excited by any of the causes which lead to congestion, or inflammation, and this ought to be counteracted by green feed, less animal food, and suppression of condiments. A little Epsom salts in the water (a teaspoonful in a gallon of water) or small doses of tincture of iron might help the cure.

Swell Head—My chickens are dying off awfully. Many of them are good sized pullets. Their heads seems to swell and they go blind and just drop off. Some of them open their mouths and stretch and act as thought something was choking them, but I cannot detect anything. They had mites but have none now. We have a good yard for them, and an alfalfa patch and some shade trees. I feed them well, and am at a loss to understand. My neighbors on either side of us have the same trouble.—Mrs. F. K.

Answer.—Your chickens have what is called "swell-head' and roup. They have either caught it from taking cold or from the lice which they used to have, or by infection from the neighbors. I think probably there is a draught in their sleeping quarters, from a crack or a knot hole or it may be wrong ventilation. Stop these up and be sure the chickens do not live or sleep in a draught. Rub their head with carbolated vaseline, and give each of those affected a quinine pill every other night for a week, and add a little poultry tonic to their food. I think as soon as you stop whatever may be the cause of their taking cold you will have no further trouble. Be sure to keep the sick fowls away from the balance of the flock.

Like a Run Around—I have quite a lot of chickens and every once in a while one of them has one eye swell up like a run around. It keeps getting larger, finally it breaks. The hen seems to get poorer all the time. I kill them as soon as I see it commencing to swell. Now I would like to ask what it is and what causes it,

and is there any cure for it?—Mrs. R. A.

Answer—Your hens have either roup or chicken pox. In either case put carbolic salve on the swelled part of their heads, and give the fowl a pill of quinine, also add a little sulphur and chopped onions to their mash at night. Be sure to keep the hens and the hennery free from lice and clean.

Swelled Eyes—What is the best cure for swelling of the eyes in half-grown chicks? They have the colony houses and are fed according to the method advised, but they seem to catch cold. It is very contagious, and seems to be running through the flock.—J. F. S.

Answer—Your chickens are taking cold, probably from a draught of some kind in their sleeping quarters. Find out the crak or hole which is causing the draught and stop it up. Put bluestone into their drinking water—a piece the size of a navy bean in one quart of water. Grease their heads with carbolated vaseline. Separate the sick from the well, for it is very infectious. Those that are sick should have a pill of quinine for three nights in succession—1 grain.

Swell Shut and Water—Will you kindly tell me the cause of sore eyes? My chickens' eyes swell shut and water. I also have turkeys; their eyes swell underneath.—Mrs. C. J. N.

Answer—Your chickens and turkeys have lice and are taking cold. They are taking cold from either sleeping in a draught or sleeping in a place that is too close and hot, so they take cold when they come out in the morning. Remedy the cause and use one of the many roup cures, and also get rid of the lice. Lice go to the eyes to drink and so spread the disease.

Something in the Throat—It would be a great favor to me if you would let me know what to do for my chickens. They are cross-breeds and run on open range, where there is plenty of good water and green alfalfa and other green grass. I have been feeding them clean new wheat; all they would eat. They are six months old, but have commenced to get sick; the first was taken sick a week ago; acted like it had something caught in the throat; opened bill and made a noise, but seems to be well now. Another commenced last night; made a noise all night like it wanted to crow; is very sick, comb very dark, droops the head slightly, eyes shut, no watery appearance and no lice or other vermin. I have examined its neck and cannot see or feel anything like diphtheria in mouth or throat; no discharge from nose; crop empty.—F. P. C., Mexico.

Answer—I think your chicks must have got and eaten some seed or burrs with beards on them, and this has formed an abscess low down in their throat, or even in the gizzard. Sometimes they stick in the throat. After a time they will get dislodged and pass through the chick without injury, but if they stick in the gizzard, blood poisoning comes on, the comb turns black and they die. When I was in Oklahoma the tarantulas sometimes bit a hen. She would fall down paralized, and act as though she were dying. I gave one drop of aconite in milk, and they always recovered under this treatment. Do you think your fowls have been stung by centipedes, etc.?

Tuburculosis—A year ago I had the nicest Black Minorcas that anybody ever laid eyes on, but, alas! one after the other I had to kill. First they get lame on one foot, then their combs get very dark, almost black on the points; their appetite is poor and they get as light as a feather, and when I cut them open their liver almost fills up their whole insides, and the whole liver is thoroughly sprinkled with little white kernels; sometimes as big as a good sized head of a pin, sometimes as large as five cents, and I attend to them so good. Now, can you tell me what disease it is and how to prevent it after this? I feed lots of green stuff, milk, meat, wheat, barley and occasionally a mash of lots of carrots.—Mrs. M. R.

Answer—I am sorry to say your Minorcas have chicken tuberculosis. You gave an accurate description of the disease, and I am very sorry to have to tell you that there is no cure for it when once it has commenced. You may be able to prevent the young ones catching it by moving them on to fresh ground, and thoroughly dis-

infecting the old yards and hen coops. I would advise you to write to the Director of the Experiment Station, University of California. Berkeley, California, and ask him to send you the Bulletin on tuberculosis of fowls. No. 161. In it you will find a full description of the disease and the steps' necessary to take to stamp it out. It is like consumption in the human family, and is considered incurable.

Ulcers—I have a hen whose feet and legs are quite swoolen, and cracked until raw. I had her setting. In about ten days I noticed her wattles and comb getting very pale, then she would not stay on her nest as she ought to. I examined her and found her legs, where the feathers start, full of mites all filled with blood. I noticed at the same time there was a small sore between her toes; thought she had picked it; changed her nest, but she would not set in the new nest. In about two weeks her feet and legs commenced to swell, and the sore commenced to get larger. Now it covers her whole foot; she also has a sore on her neck that gets larger. It looks like the same thing; both sores are getting larger. I have used vaseline, carbolic salve, and bathed and bandaged her feet, but to no purpose. Can you please help me? Give the cause, and if contagious.

Can you tell what is the matter with my young chickens three months old? They seem droopy. My chickens all have plenty of range, green feed and clean water.—Mrs. P. M.

Answer—Your hen should be killed to put her out of her misery. From your description of the trouble, she has ulcers, which may be turbuculous, and if so, they are contagious, and there is no cure for them, so the best thing you can do is to kill her, and burn her body. Do not bury it.

I think your chickens are droopy from having mites and perhaps lice also. You should spray the houses, or whitewash them frequently, and once a week dust all the chickens with some good insect powder. If you are not careful your little chicks may be bitten by the same lice and mites that were on the old hen and they many catch the same disease.

Vertigo—Being an interested reader of your question department, I take the liberty of asking you about my little chicks. They have a queer disease that I never saw before. They commence to hold their heads to one side, keep twisting their necks until they fall down and roll over and seem in a kind of fit, and then jump up; seem better for a while and then go through with the same performance until they die. They peep as if in pain. I have lost several. I feed cornbread and sour milk curd and they run in the orchard. Do you know what it is and is there a cure for it? They have no vermin.—Mrs. R. B. L.

Answer: Your chickens have vertigo. This is usually caused by acute indigestion, from wrong feeding, from sunstroke, from intestinal worms, from poison or from lice. Overcrowding the chicks also has a tendency to bring it on. I have known of several cases similar to yours from the chicks having eaten putrid meat. The best treatment is a little Epsom salts in the water, about a teaspoonful to a pint of water. Give this as their drinking water. Give plenty of fresh clean water and green food. If you think it is worms, put a teaspoonful of turpentine in a quart of the drinking water or mix their mash with it and give it also to them to drink. This will kill the worms. If you think it is from poison, give each chick a pill of asafoetida, about a two-grain pill or even smaller if the chickens are very small.

Wind in Crop—Will you please tell me the cause and remedy of my little chicks, from three to four weeks old, having a gas gather in their crops? When the crop is pressed, wind comes from the mouth and they stand around and gasp, but otherwise do not look droopy. They eat well, but in three or four days die. I lost quite a number last spring, almost every case being fatal. I have a hen with young ones and I would like to raise them without this trouble.—B. C.

Answer: The wind in the crop comes from indigestion. Indigestion comes from lice, colds, dirty water, and chief of all from wet mashes or from wrongly balanced food, and lack of hard, sharp grit to grind the food. I do not think the chicks with the hen, if she is allowed free range, will get it, but if there are any symptoms of it, put some lime water into the drinking water and give them pounded up charcoal. Give them also sweet skim milk to drink as well as water and plenty of nice, crisp lettuce to eat. I am sure if you keep them quite clean, feed clean dry chick feed with plenty of green lettuce, grass or clover, cut up fine, you will not have any wind on the stomach with your chicks. A little bicarbonate of soda in the drinking water will sometimes help, but prevention is the best cure.—R. W.

White Comb—My fine Orpington rooster is developing a peculiar disease. A few months ago he was in the pink of perfection, but his comb has become all covered with white spots, as though he had dandruff, and it spoils his appearance. I feed your well proportioned mash, wheat, alfalfa, crushed green bone, corn, lettuce and cabbage; a mash every morning and corn or wheat for the evening meal. He is vigorous and active, the only trouble being with his comb. If you will kindly tell me how to treat him for this trouble, it will be highly appreciated.—E. R. T.

Answer: Your rooster has what is called "White comb." It usually comes from close air in the hennery and a total absence of all green food. It is a contagious disease and may be imparted from bird to bird, probably also from mice, rats, cats and dogs to birds. Young birds appear to be more susceptible to this disease than old ones. Put carbolated vaseline on the comb and, in the drinking water, use twelve tablets of nux vomica and sulphur comp. 2X to each pint of drinking water. Continue the treatment until cured.

LICE, MITES, TICKS AND WORMS

Body Lice—I have about 100 White Leghorn chickens and I find that they have a large body louse, large yellow ones; what can I do to get rid of them? I think they are keeping my chickens from laying as they should.—Mrs. B. W.

Answer—Paint the bottom of a box or barrel with a good lice killer; put a little straw in to keep the paint from the feathers, then put the chickens in and cover them three hours. Then examine the hens and pull out all the feathers that have nits (lice eggs) on them, putting the feathers into a little can of coal oil. Then dust the hens with a good insecticide once a week or until you are sure all the lice are dead. Be careful to give the hens a spot of ground, well spaded up, mellow and a little damp. They will bathe in this and usually keep themselves clean.

Dipping Hens—Would you be so kind as to let me know about dipping hens, etc? I have a flock of some five or six hundred. I notice some of them have lice and bunches of nits on their feathers. Whenever I have caught a hen I have greased her well but this would take too long to go though the bunch. Is there any dip that would be strong enough, and do no harm to the birds, that would kill the nits with one dipping?—W. L.

Answer—Lice are supposed to hatch out the nits every five days and when but a few days' old commence to lay again and so keep on breeding indefinitely. Dr. Salmon says it has been estimated that the second generation from a single louse may number 2500 individuals and the third generations may reach the enormous sum of 125,000 and all of these may be produced in the course of 8 weeks. I do not know of any dip that will kill the nits with one dipping. Dr. Salmon recommends a dip of one per cent carbolic acid solution, or using creolin as it is equally efficacious in killing insects and is less poison to the birds. It is used in the strength of two and a half mixed with a gallon of water. I have used very successfully in the summer time when the weather is warm the kerosene emulsion made as follows: Dissolve one bar of soap or one pound of soap powder in a gallon of boiling water; add to it a gallon of coal oil and a pint of crude carbolic acid; churn for twenty minutes or until you wish to use it. Take one quart of this top solution and add it to 9 quarts of water. Dip the hens into this, being careful not to allow any of it to go into their eyes or mouth, but thoroughly wet every feather to the skin. This will kill every living louse and if repeated, in about five days will probably kill those that are hatched out in the meantime and prevent their laying any more nits. Tobacco water has also been strongly recommended as a dip, and chloro-naphtholium used as directed on the bottle.

The Sand Flea—How can I rid my chickens from a small insect known here as the sand flea? I have tried coal-oil mixed with lard without effect. The hens scratch their heads so they become sore and some have died; others have had to be killed.—Mrs. F. A. F.

Answer—Those fleas are very hard to get rid of. Spray the henneries well with either the kerosene emulsion or good hot salt water, and while the ground is still wet, scatter on it, air-slaked lime. Those hens that have sore heads, should have carbolated salve put on them, after swabbing them off with corrosive sublimate. This will kill the fleas and cure the sores. Be careful not to let any of the corrosive sublimate get into the eyes or mouth of the fowls.

Stick Tight Fleas—We have noticed a tick or louse on a few of our chickens and have discovered some of the insects on the perches. They resemble small black beads and are firmly embedded in the skin. On some of the fowls we have used for the table we noticed a few red blotches on the skin. We would like to know how to get rid of the insects, particularly how to get them out of the hen-house.—An Inquirer.

Answer—You have the stick tight fleas in your hennery. They are very hard to get rid of, being in some places a perfect pest. A friend of mine lost 500 out of 700 chickens last fall from this. I told him to spray very thoroughly with salt and water and he purchased 600 lbs. of salt, scattered it all over the hennery and yards and then turned the hose on them for several days in succession. He tells me now there is not a stick tight flea on the place. I advised him to get some corrisive sublimate diluted with alcohol at the drug store, take on old tooth brush and carefully apply with it the corrosive sublimate on any fleas he might see on the chickens, being careful not to allow any of the solution to get into the chickens' eyes (it would blind them) or into their mouths as it is very poisonous. You can paint the perches with this; it will kill everything it touches.

Head Lice—This time I write in desperation, hoping you may be able to give me a remedy. It is head lice I am fighting, and after working for almost five months, I am as far off from being rid of them as at first. I have done everything that I have ever heard of. I still find they have head lice and red mites besides. I hope no other beginner has had the trials I have had.—Mrs. W. F. K.

Answer—The red mites live in the houses or coops, except when they are feeding off the chickens, usually at night. The cure for them is to spray the coops thoroughly and constantly. You can keep them out of the coops by spraying once every three weeks, but if they once get in, you will have to spray twice a week until you get entirely rid of them, then once every three weeks, to keep rid to them. The head lice live on the heads of the chickens. They lay two or three white silvery nits (eggs) at the root of the feather. The eggs hatch in about five days after they are laid by the lice, consequently to completely destroy them, you should treat the chickens that have them, at least once a week. The best way I know of is to take an old toothbrush, a bowl with nice hot soapsuds in it and a few drops of the best carbolic acid; brush the chicken's head with this, being sure to touch all the lice and mites. This, I know, is an excellent remedy for I have tried it. Another given by a friend of mine is, get the druggist to mix some corrosive sublimate with the best pure alcohol, take the tooth-brush and brush the chickens' heads with this, being very careful not to let any of this get into the eyes (or it will blind them) or into the mouth, as it is very poisonous. This will not only kill the head lice and their nits, but it will also kill stick tight fleas, ticks, and any insects. It is very difficult when

once the pests get into henneries or on chickens, to get rid of them. It is far easier to keep the enemy out by constant and thorough cleaning at frequent intervals, especially in the summer time. I find using tobacco stems for making the nests of setting hens, a good preventative; besides this, I see that all the fowls have good dust baths in damp and mellow earth.

Hump Themselves—I will have to come to you with my sick chickens. It seems to be chicken raisers' only refuge. I have lost several half-grown and whole-grown. They kind of hump themselves all together, do not care to eat; do not stir around. I never noticed any bowel trouble; it looks to me like their heads turned dark; live several days. What shall I do?—L. H. E.

Answer—It is very difficult to diagnose a case like yours with so little information about it, but from your description of the chickens humping themselves and appearing sleepy, I think they have worms. You should open one and make a thorough examination; then you will know what really is the matter. If it is worms, give them thirty drops of turpentine in a pint of water. Let them have no other water to drink for a week and I think it will cure them. Possibly they may be taking cold and very probably may have lice. Examine them and dust them, and try to discover what is giving them cold. Give them a little poultry tonic and follow my directions for the general care of fowls.

Mites—We are fighting mites, but apparently with no success. We hired a man who makes poultry ranch spraying a business. We paid him $10 and he guaranteed to rid the place of the pests, but they are worse than ever. He uses lime, sulphur and carbolic acid. Is there any way corrosive sublimate could be used as a spray, and would it be safe for the hens in the houses? How long would the hens need to be kept out after the spraying was done? Am having the worst possible luck with my chickens. Have probably hatched 550 chickens this year and have less than 200 now. When a week to ten days old they begin to droop, refuse to eat, and starve to death. What is the matter? No bowel trouble; no cold; no lice, or only a few. Does cholera ever attack such young chickens, and if cholera, would they not have bowel trouble? Would greatly appreciate an immediate answer, as the mites get all over me and drive me nearly frantic. In putting down eggs in water-glass; must they be infertile eggs? —Perplexed.

Answer—The thing that is killing your little chickens is not cholera, otherwise they would have bowel trouble; it is only the swarms of mites. If they drive you nearly frantic, think how the chicks must suffer. The mites simply drain the life out of them. The corrosive sublimate can be put on with a spray, but it is dangerous to do so, as if it splatters into the person's eyes who is spraying, it may blind him for life. One pound of this costs $1.25 and that is sufficient to made 120 gallons of the solution. As it takes some time to dissolve in water, it is usual to dissolve it in alcohol. I have used it dissolved in alcohol to paint henneries and nest boxes and it will destroy all insect life. You must turn the hens out of your henneries for several hours, or until the walls are dry. The eggs need not necessarily be infertile, although it is better to have infertile eggs, but above all things, they must be absolutely fresh. One stale egg may spoil the whole crock full.

Ticks—We have a pest the like I never saw or heard of, so I come to you to know what they are, and if they are common in California; they occupy every crack and crevice in the hen house, never stay on the chickens during the day, so far as I can find out, but just simply bite them at night until the flesh looks skinned in places. Now for a description of them. They are dark grey, flat and have plenty of legs. I have seen them as large as the thumb nail.— Mrs. M. R.

Answer—The pest you speak of are ticks. They are indeed a terrible pest and you cannot possibly succeed with chickens unless you get rid of them. They are very common in some parts of California. I have had specimens sent me from widely different places, and at a large farmers' or grange convention the subject of getting rid of them was discussed by about sixty farmers and professors of the California University and no definite remedy was found. However, some of those who have applied to me have by my advice used corrosive sublimate—eight ounces to twenty gallons of water. Painting the henneries with this was the only rem-

edy found to be effective. Kerosene, distilate, whitewash and everything else proved useless, but the corrosive sublimate did kill them. Care must be used in handling this, for it is poisonous, and if any gets into the eyes of the worker, it may blind him. These ticks are very much like bedbugs, and I have seen them almost as large as and resembling a watermelon seed, after feasting on the chickens; before that some of them were as thin as a piece of paper, lying flat between the shingles on the roof where it was almost impossible to get at them. The only remedy is the persistent use of the corrosive sublimate.

From Wild Birds—Some years ago my fowls became afflicted with a round worm, also tape worms, and in one article you mentioned several remedies such as santoine, turpentine and tincture of male fern. I dug up the yards and seeded to green feed but all to no purpose; it has practically driven me out of business. Last Spring I invested in some outside stock (just hatched baby chicks), but they also became infested although they were on new land. However, I managed to keep down those pests by occasionally dosing the hens with the above mentioned medicines. We do not feed anything unclean to our fowls and it always has been a puzzle to me where such worms came from.

A few days ago our house-cat brought home a small bird, which she began to devour on the house porch, but leaving the intestines, out of which crowled two good sized round worms such as fowls have. As we live in the woods, do you think this has anything to do with it? I am almost afraid to start my incubators this season, as it may only result in future failure.—W. E. B.

Answer—Your fowls undoubtedly get the worms as the wild birds do, from the droppings or eggs of worms from the other birds. By the persistent use of turpentine, using thirty drops in a quart of water, or mixing it in that proportion in the food, for a week at a time, you can get rid of them. Also disinfect the ground. The only thing that I can see is for you to keep up this treatment, for a week every two months, giving turpentine either in the food or water. I would not be discouraged because that is a sure remedy and by watching and noticing the droppings, you need not fail in rearing the chickens.

From Pigeons—My chickens' gizzards are affected by red worms about the size of a pin. All the stock I raised last year seem affected although the eggs came from different places. I have the Brown Leghorns, Brahmas and R. I. Reds. I feed all the various grains, plenty of greens and good meat and bone. The only thing you recommend that I have not fed is charcoal, still as chicks they got it in the chick feed. I have given them turpentine in food and water at various times and it seemed to have the desired result, but today I learned different, the gizzard is penetrated and has a sore spot caused by these worms. All the stock in different yards are affected.

I get plenty of eggs and the chickens look good, combs nice and red, nevertheless I find them all affected the same way.—Mrs. G. S. L.

Answer—I have been through the same trouble myself and so can help you. The difficulty is to find the source. I found out that my chickens were getting the worms or the eggs of the worms from neighboring pigeons. The droppings of the pigeons contained the eggs of the worms and in a short time the droppings of the chickens also had them and the other chickens ate them and so on they kept increasing. First of all I gave the chickens the turpentine which I recommended to you. A teaspoonful in a quart of water. Mix the food with that water, also put a teaspoonful in a quart of the drinking water and allow no other water for drinking. Keep this treatment up for a week. Meanwhile clean up the yards by having them either ploughed under or dug up and a crop of some kind planted, something that will grow quickly, such as wheat or barley, and as far as possible destroy the birds that are bringing you the trouble, for I cannot but think it must be pigeons or some other wild birds. The worms will kill the young chickens, but they do not always kill the older fowls. Sometimes the worms come from unclean or spoiled food, from "webby" grains and bad animal food. You will have to discover for yourself where they are getting the worms from and cut off the source of supply.

Intestinal Worms—I wish a little information and advice in regard to a valuable Buff Orpington cockerel I own. He has become mopy and goes away under the trees by himself, and has lost over half of his weight in a month. He eats like a horse, though, of everything I give my hens, but shakes his head an awful lot, as though something was wrong. I looked in his throat and it looks all right. He has changed in color from a light buff to a very dark red since acting unwell, and has grown to be a homely, dopey bird, from a real beautiful lively one a shore time ago.—M. J. Q.

Answer—I think your Buff Orpington Cockerel has intestinal worms. You had better give him 25 drops of spirits of turpentine on a lump of bread, or in a spoonful of water, and follow that immediately with two teaspoonfuls of castor oil. Keep him shut up so you can watch the droppings and remove and burn or bury them deeply. If you do not find worms in his droppings, give him ten drops of tincture of male-fern on a lump of sugar, followed in an hour by a dose of castor oil. This is for tape-worms. Both the remedies should be given after twelve hours or more fasting.

Bantam Affected—I have a little hen, bantam, in whose droppings I noticed what look like wormns. She is thin and looks like she has catarrh. Can you help her? Also a Plymouth Rock rooster who has a film over his eyes and sleeps all day, begins to take exercise about sun down; appetite fair. I feed every variety of chicken food alternating, and keep shells, charcoal and green food, and they are not fenced in.—J. L.

Answer—Your little bantam hen undoubtedly has worms, as you see them in her droppings. Your Plymouth Rock male bird also has them, for sleepiness is one of the chief symptoms of worms in the intestines. The best cure I know is turpentine; ten drops in a teaspoonful of castor oil, after the chickens have fasted twenty-four hours.

If you have other chickens and think they may have worms, you had better give the whole flock some turpentine in their drinking water. Thirty drops of turpentine to a pint of water. Do not let them have any water without turpentine in it for a week.

Several Kinds—I am in despair and it is lice, lice, lice. We have Brown Leghorns and as they will not sit we borrowed a setting hen and she only stayed with us long enough to give our hens a supply of grey head-lice. When we discovered them we went to work with a lice killer, sprayed the coops, ground and nests, put the chickens in a box and left them three hours. We also used crude oil, poured gallons on the ground, painted nests, roosts, etc., but still the lice stayed on the hens' heads. Last week we bought six Buff Orpingtons; yesterday we found they were alive with body-lice, yellow lice, especially around the vent; there were thousands; then we examined the Leghorns, found they were infected also. What shall we do? Do you think it would hurt them to wash them now with the kerosene emulsion? Am afraid it might give them a cold.—Mrs. C. S. B.

Answer—What I should do were I in your place would be to get some buhach powder, rub it well into the chikens' heads for the head lice, and well into the fluff under the wings and on the backs for the body lice, then put the hens, six or a dozen at a time, into a large size dry-goods box, at the bottom of which is a newspaper thoroughly painted with a good lice killer; cover the top of the box with a carpet and leave them in for three hours, then look them over thoroughly and pull out every feather that has nits on it. The nits hatch out about every five days, so in a week's time, look the hens over again, powder them again, and again put them into the box painted with the lice killer. Two applications should cure them. After this, once a month, at night, powder them with buhach and look them over occasionally, and if necessary, go through the performance again. You can paint the roosts with

lice killer, but do not put any in the nests, for it will not only flavor the eggs but will kill the germs and make the eggs unhatchable. The best thing to use for the nests is a kettleful of boiling water with a large handful of salt added to it, or scalding soapsuds, putting in fresh straw, or better still, making the nests of tobacco stems. You can get these for 25 cents a gunny-sack full.

Do not risk washing the hens except in the hottest summer weather.

Spray for Houses and Dip for Hens—Last summer I found a recipe in one of your articles for spraying henhouses. I used it to good advantage but have misplaced the recipe and cannot remember the mixture exactly. It was composed of coal oil, carbolic acid and soap, with a certain proportion of water. If you will kindly send it to me I will appreciate it.—C. W.

Answer—I gladly send you the recipe which is excellent. I have used it for ten years or more. It will kill fleas, lice, mites or any insect pests in the henneries. It will also thoroughly disinfect the premises from infectious diseases and if used for a dip for hens in warm, sunny weather, will rid them of lice and will assist the moult:

Dissolve one pound of hard soap (or soap powder) in one gallon of boiling water, remove from the fire and add immediately one gallon of kerosene and one pint of crude carbolic acid. Churn or agitate violently for twenty minutes or until you want to use it. If the oil and water separate on standing, then the soap was not caustic enough. Add to this ten gallons of water.

I keep the stock solution on hand, dip out a quart and add to it ten quarts of water and use it for spraying the houses once every three weeks in summer and every month in winter. Putting it on hot in summer and slopping it well into dark and dusty corners will kill fleas, which are exceedingly troublesome on sandy soil in this part of the country.

FEEDING IN GENERAL

Feeding System—I am not perfectly satisfied with my feeding system and I follow yours on the food question. I note that you advise dried blood and other food dried in the oven, green cut bone and bone meal. Would you advise boiled liver, lungs and scraps instead of prepared meat scraps? Are ground clam shells good in place of cut bone? Could there be any danger from feeding too much ground shell? Should gravel be furnished to chickens to pick from?—D. F.

Answer: Boiled liver and lungs chopped fine are excellent for fowls. I prefer them to prepared meat scraps. They must be fed while fresh as spoiled meat may poison the fowls. Clam shells cannot take the place of cut bone. Crushed oyster and clam shells contain lime which is very good for making egg shell. There is no danger of the hens eating too much of this. Gravel or grit should always be furnished to chickens.

Animal Food—I would like to know what you would suggest in the way of animal food for my Plymouth Rocks. Have only eleven hens; they have free range and running water. Some are laying, but I do not give any meat or blood meal. As we only go to town about once per month it is rather hard to bring out anything in that line and keep it fresh. I feed rolled barley, and lately have given them a mash of shorts or bran, and some cayenne mixed in it. Hope you will suggest something, also how would you feed them this winter? They have plenty of grass on their range.—Subscriber.

Answer: Fresh meat is best for hens. Can you get rabbits or squirrels or gophers for them? If not and you cannot obtain any of the good egg foods from the supply stores, in your part of the country, you might try the following. which I have proved to be excellent: Take ten pounds bone meal, ten pounds dried blood, five pounds linseed meal, two pounds sulpher, two pounds powdered charcoal, one-half pound cayenne pepper and one-half pound salt. Mix and keep. Put a half pint in your mash every day for twenty hens. When you feed this, feed no meat scraps and do not salt the mash. You will get the mixture right if you remember that the combined weight of the ingredients is thirty pounds. This is simple and cheap.

Bad Meat—I had twelve laying hens, they averaged seven eggs a day, were healthy and never were sick until I bought five cents worth of green ground bone from a wagon that passes my door. It was wet and slimy, and smelled, but he said it was all right. I gave it to the chickens at noon; fed them nothing else then. At four o'clock I went out and found two dying and six more droopy and by eight that night had lost eight. Next day two large Buff Orpington hens died. I looked for some of your remedies giving asafoetida pills and the soda you spoke of in the water. I showed the bones to the butcher and he said he never heard of such a thing as spoiled meat poisoning chickens. He sold it when it smelled like that all the time.—Mrs. D. M.

Answer: That meat poisoned your chickens evidently. It is called ptomaine poisoning. Butchers sometimes put formaline or some preservative on the meat which has a very poisonous effect on chickens, but yours were undoubtedly poisoned by the putrid meat. You had better not buy any ground bone unless it is quite fresh.

Blood Meal—Will you pease tell me how much blood meal to put into the mash for thirteen chickens, or in other words. what proportion for each hen?—L. S.

Answer—Half an ounce per hen every day at this spring season of the year is about what they need of blood meal mixed in the mash. Weigh out enough for the thirteen hens and measure that in a cup or by a spoon, then you will know how much by measure.

Analysis of Barley—Would you kindly let me know what the chemical analysis of barley is, in its composition? It is sometimes more convenient to use it than other grains, and I want to know its value as a food for poultry.—F. I. D. W.

Answer—The chemical analysis of principal grains and poultry foods in California are given in full by Prof. M. E. Jaffa in his most useful Bulletin No. 164, published by the College of Agriculture, Berkeley, California. (This Bulletin is free to residents of this State.) Prof. Jaffa gives as the analysis of rolled barley: Digestible nutriments in 100 lbs., protein 9.3, carbo-hydrates 59.5, fat 2.2, nutritive ratio 1. 6.9. Barley has about the same nutritive ratio as plump wheat. I would advise you to send to the director of the Exeprimental Station, University of California, for the Bulletin, 164, as by means of this little book you will find it easy to balance all your rations.

Balanced Ration—Will you please send me a balanced ration for White Leghorns? I have, perhaps, 125 hens, but I get only six or eight eggs. I will have to buy all the food. I have a half acre patch of alfalfa and plenty of good water; keep it running into a slatted trough all the time. Have been using white-wash with crude carbolic acid and kerosene for mites and other vermin. My hens roost in trees and are looking fine; plumage white and glossy, combs red, but no eggs. I have houses with good roosts in which I put my chickens from brooders, but in a short time they go to the trees to roost. My hens have the range of forty acres if they wish but it is salt grass pasture.—A. C.

Answer—If you will tell me what grains and what feed of all kinds for fowls you can get the cheapest in your part of the country, I will gladly write you out a balanced ration, composed of those grains; this will be your cheapest plan. From what you say of your hens' beautiful plumage I think they must be nearly through the moult. Leghorns take longer to moult on account of their close, hard plumage than the American or Asiatic breeds, and they do not lay plentifully until the new feathers are fully matured. To hasten their laying, give them more animal food than you are doing; increase it carefully, but feed them more. The onions will have to be chopped and you will have to feed them rather sparingly when the hens are laying, or they may flavor the eggs. Onions are very good for growing chicks, ducks and turkeys, but must be fed sparingly to laying

hens. Hens that roost in the trees are apt to get damp in the winter time, and will therefore not lay as many eggs as if they were kept dry at night. They will also need more food, consequently in this climate it is best for them to roost under a roof.

Beet Tops—Will you kindly tell me if beet tops are a good green food for ducks? Also for fowls and turkeys? Are they as nourishing as alfalfa? My hens are not laying well. The eggs have suddenly dropped off and I did not know but what the cause might be beet tops.—J. S. Y.

Answer—In September one is glad to get anything green for the fowls, ducks, geese or turkeys, to eat. Almost anything green is better than nothing, but alfalfa contains more protein than any other green food except white clover. The per cent of protein in white clover is 15.7 and in alfalfa 14.30 while in beet tops it is only 1.3. By this you will see that alfalfa is worth about 14 times as much as beet tops. There is about as much protein in alfalfa as in wheat bran. You complain that your hens do not lay. I think probably they are moulting. You cannot expect hens to lay all the time without taking a rest.

Dry Hopper Method—I write you regarding the dry hopper method of feeding. How much space do you leave at the bottom for the feed to come through, and how wide do you leave the space for the chickens to eat out of? We made one but its not a success, for the box is bloody from their combs hitting against it. They stand and eat all the time and do not go and drink as you say yours do.—D. S. M.

Answer—I had the same experience with hoppers injuring the combs of the fowls, and now I make my hoppers like those used at the Maine Experiment Station, simply a box with a roof over it. The box is twenty-four inches long and eleven inches wide. The sides are cut like a gable, the highest point being sixteen inches high. The gable roof keeps the food dry and the hens waste scarcely any of it. The roof lifts off or can be slid back to fill it.

Dry Mash—Will you kindly inform me as to the best method of feeding alfalfa meal to hens and pullets? I use hopper constantly filled with dry mash consisting of bran, shorts, feed meal and beef scraps, accessible at all times, and would much prefer adding the calfalfa to this. Or would you advise soaking it in water and feeding it separately? The fowls get grain twice a day, and now if I add the alfalfa to the mash what proportion shall I make it? Also, is it as well to add the charcoal, two or three per cent, to the mash or feed separately? I wish to simplify the routine work as much as possible.—Mrs. O. K.

Answer—I advocate adding the calfalfa meal to the dry mash. It would make a very good ration to simply add one part of calfalfa meal to your present mash, making it one part each of bran, shorts, feed meal, beef scraps and calfalfa meal. I feed this with excellent results, but at first the hens did not like the calfalfa, so I only added one iron spoonful, increasing the dose every day, adding one more spoonful until, within a month, they were having the right proportion. You can mix the charcoal in the same way, but I prefer to keep it separate with the grit and the crushed shell.

Exercise for Fowls—I was greatly interested in an article of yours on feeding. You say, give a hen a chance to work and no matter how fat, etc. Now what interests me most to know is just how you manage to give them plenty of work in a limited space. We, who occupy only a village lot, will be greatly helped if you will tell us how to keep hens busy in such limited quarters.—G. P. C.

Answer—To keep hens busy, give them what is called a "scratching pen." Put a 12-inch board across one corner of your lot and fill that full of good wheat straw or hay; scatter all the grain you feed in that, and the hens will work all day digging out the grain; every grain they scratch out they will bury two and so will keep up the exercise. If you are feeding the hopper method, put the hopper at one end of the pen and the water vessel at the other end; this will give them the exercise of walking back and forth. You can also hang up a cabbage for them to

jump at, but scratching is the natural and best exercise for developing the egg organs.

Ration for Twelve Hens—I take great pleasure in reading your articles. One thing I have failed to find and that is a good balanced ration; many writers say, feed a good balanced ration, but few of us new beginners know what a good balanced ration is. We are just as apt to overfeed as to under-feed. Would you kindly give me formula for a good egg ration? In giving ration, kindly state quantities of each kind of feed used in ration, amount to be fed to twelve hens, whether to be fed wet or dry, morning or night; also amount of grain for twelve hens; in other words, a full day's egg ration for twelve hens; when to feed, how to feed and quantity for daily ration. I have some White Plymouth Rocks, over eight months old, large and well developed, but only two of them have commenced to lay. I feed morning mash of 2 parts bran, 1 shorts, one barley meal, one cornmeal, one alfalfa meal, ½ blood meal. Wheat at night, about 1½ pints for twelve hens; good clean yards and houses; fresh cut kale at noon.—W. S. F.

Answer—The ration you are now feeding is a very good one, but at this time of the year (early spring), I would advise you to double the amount of blood-meal in the mash. I would feed the mash perfectly dry, without moistening it in the least, in the morning; the green feed at noon, and the wheat at night, or I would reverse it, feeding the wheat in the scratching pen in the morning, green food at noon, and the mash slightly dampened with table scraps you may have, at night, giving the hens at their supper time, what they will eat up clean. Pullets that are ready to lay will sometimes retain their eggs if they do not have comfortable nests; also sometimes they require a slight shock or stimulants to start them laying. I find chili pepper seeds excellent for starting the laying, or failing to get this, a teaspoonful of red pepper three times a week for a dozen hens, will often start them laying. The ration you are feeding, if you add more blood meal (or animal food) is a well balanced ration for eggs.

Tomatoes—Do tomatoes tend to make the hens quit laying?—J. W.

Answer—Tomatoes will not do the hens any harm unless fed in very large quantities. There is not much nourishment to them and consequently they will not improve the laying qualities; otherwise a reasonable amount will benefit the hens.

Formula for Feeding—Your formula for feeding—two parts bran, one part cornmeal, one part alfalfa meal, one part shorts, one part beef-scraps —is the simplest I have ever seen, so shall try it.

1. Will the same formula hold good with hens with free range but no green food?

2. In case they have access to fresh alfalfa hay, would it be necessary to use the alfalfa meal?

3. Could I substitute shorts or middlings for the meal in case they are cheaper, and if so, in what proportion?

4. Does the balanced ration keep up the egg yield during moulting or is it necessary to add oil-meal, or some similar meal during that period? —Mrs. G. H. G.

Answer—The same formula is good for hens with no green food, but it is much better to give them green food, or roots, beets, turnips, carrots, pumpkins, or some succulent vegetable if possible.

2. No, not absolutely necessary, but I always continue the alfalfa meal so the hens may not forget the taste of it, as it is sometimes difficult to break them into the habit of eating it.

3. You could not substitute shorts or middlings for it.

4. During the moult, add oil-meal or linseed meal, about one-fourth of one part, to the feed. This ripens the feathers, makes them fall out easier and grow more quickly.

For Young and Old Stock—I am very much interested in your articles and would like to ask you for a little advice. Being away from home all day, I have to feed in the morning enough to do all day. This I can manage for the old stock by feeding scratch food in the litter and dry mash in hoppers. But how can I manage the growing stock? Please give a formula for dry feed. Do you consider the scratch food sold by the poultry houses good food for the young stock? My chicks will not eat the baby chick food after a week or ten days. I also give them lawn clippings or lettuce every evening.

Is a handful of scratch feed to the hen once a day enough where they have the dry mash and table scraps? Is cracked corn good food to feed alone to young stock? I have Rhode Island Reds.—R. L. P.

Answer—Your questions relate principally to the feeding of the young stock, and you do not say whether you want to keep them for fattening for the table or for future egg layers. There is of course a difference in the way of feeling, or rather in the quality of the food to be given to them. However, I will tell you the way I feed for egg laying. As soon as I think the little chicks will eat whole wheat I add it to the baby chick feed, a small quantity. If they pick it up quickly I add more each day and in a few days I give also some kaffir corn or finely cracked corn. It should be finely cracked as it is difficult of digestion. When it is too long in digesting the corn ferments in the gizzard and that gives the chick diarrhoea, which often proves fatal. We never want to overtax the digestion of a chick, so I give corn carefully. This applies to the last question in your letter—it is not good to feed corn alone. It has been clearly proven that chicks do better grow more quickly and mature earlier if they can have a great variety of seeds to eat. This is the reason we prefer to buy the chickfeed already mixed from the supply houses. They have greater facilities for getting a variety of grains than we have.

When the young stock is old enough to eat the wheat and kaffir corn they can be fed as you do the old hens, only remember to give them nice, clean litter to scratch in. It will need renewing oftener than that of the old hens, for if it gets foul and they pick up some of their own droppings you will soon have a set of sick chickens. Feed the grains in the scratching pen to the little chicks and also give them in a hopper bran, alfalfa meal, corn meal, ground bone and either granulated milk or dried blood in equal proportions. The little chicks will prefer the grains in the scratching pen and eat those the first, which is just what they want, but if they are hungry they will go to the hopper. Most of the poultry supply houses now make an excellent scratch feed; they realize the need of it and are able to mix it scientifically. I always buy from them and if I think there is too much corn and that my fowls will become too fat, I say "Please economize the corn." You will find most of the poultry supply houses willing to mix the scratch food just as you want it. You are feeding the mature stock all right. One handful of the scratch food in the litter is about right for the hens. The green food is quite important, the lawn clippings should be of clover or as much clover as possible, for the blue grass becomes so hard and stiff as the summer continues that there is not much nourishment in it and the hens will not eat it. Lettuce is good but is sometimes quite expensive and difficult to get, but there is another green food that has been found excellent and is within the reach of any one. This is sprouted oats. Take half a bucket of oats, pour warm water on them and leave them covered all night, then spread them in boxes. Any box will do. Have the oats about three inches deep and keep them wet. In four or five days there will be a mass of tender green sprouts. The hens will eat eagerly of this. A friend of mine has also done this with barley for many years with great success. This green food is as good for the young stock as for the old.

In your place I would feed as you do, throwing scratch food (a handful to each fowl) in the litter in the early morning, keeping the dry mash in the hopper, and feed the green food in the evening. Some of it may be left till morning, but will not wilt much and they will eat it the first thing. Be sure they have plenty of water and have it shaded from the sun, either in a box on its side or in some sort of shelter.

Mixing Foods—I want to ask you if there is any good reason for not mixing foods at the same meal. Prof. Jaffa of the U. C. said on one occasion that it was best not to mix foods —in feeding wheat, to feed that alone; the same of barley or of corn. Make either an entire meal. I have observed in feeding my chickens that they seem to enjoy a variety of grains

fed together. Which method would you think best?

I am feeding rolled barley dry. Would you think it better to soak it? I give the mash at noon, dry, and green feed morning and evening. The fowls seem to like the green feed better at those times than at noon.

Would you set eggs from well grown White Minorca pullets that are now nearly eight months old? They are now with a rooster of the same age; or if not now, would it be safe to set them after they are nine months old?—G. S. H.

Answer—The reason Professor Jaffa thinks it best not to mix foods is because some hens will pick out all of a certain grain in a greedy manner, and by giving only one grain at a time, they are forced to eat what he chooses to give them. I would not venture to differ from so learned a man, but like you, I notice my hens enjoy a variety, so I give it them, and for the little chicks, I am positive a great variety is by far the best for them. I found that the hens enjoyed an occasional feed of soaked barley, so I poured scalding water over a few pailsful of barley, covering it with gunny sacks to keep in the steam and when thoroughly soaked, fed it to the hens.

I would not set eggs from such young pullets. I would wait until they are nine or ten months of age; especially as they are mated with a cockerel of their own age. The offspring of immature fowls is often weakly and delicate. I have found it much more satisfactory to hatch only from two-year-old birds. Then you have the foundation of a vigorous flock of fowls, and I never hatch from Mediterraneans of less than a year. It really pays better and is much less anxious work having only vigorous chickens, chickens that cannot help but grow and develop as we want them.

How Much to Feed—Can you tell me how much feed an average Leghorn should have in weight with a free range of two acres of alfalfa? Is green ground bone necessary all the year round or only in the winter? My hens will not lay and I may not be feeding right, although a few Wyandottes I have are too fat, but they get exactly the same food as the Leghorns. I have 72 hens and only got 12 eggs yesterday. Am not satisfied with the results and desire to have them do better.

Answer—An average Leghorn hen should have in weight for every pound weight of hen an ounce of food. As Leghorns weigh about five popnds each, they would require about five ounces of food each per day. Animal food of some kind is necessary for hens if you want them to lay. If you can give them milk in large quantities, that will give them all the animal food necessary. Green ground bone is, of course, the best food, but it is very difficult to keep it fresh and sweet in the summer time, therefore dried bone and dried blood, or beef scrap or milk must take the place. A hen requires about half an ounce of green ground bone every day or of the dry stuff (bone and blood) half an ounce every other day. If the fowls have plenty of green food and are not laying well give them more animal food. Perhaps your Leghorns are two years old, in which case you had better get younger fowls, as their days of greatest usefulness are over.

Feeding for Market—What shall we feed young cockerels to prepare them for market?

Our turkey hens are still laying. Will they lay next year in time for hatching season, say January of February? Of course I do not expect you could tell exactly what a turkey hen would do, but would like your idea of it. If I thought they would not lay before March, I would rather sell them . What would you advise?— S. L. J.

Answer—For fattening your cockerels, coop them in a small place, so they will not exercise. Feed them three times a day a mash composed of one part each of corn meal (feed meal), bran and rolled oats, with a little charcoal, and mix it with milk, if possible. Take away the food in fifteen minutes, leaving only water and grit before them; give them all they will eat of this, and in from two to three weeks they will be delicious, fat and juicy. The last week add five per cent linseed or cotton seed meal.

Your turkeys that are laying now will moult late and probably not commence to lay again before March or April, although as you say, one can-

not be very certain what a turkey hen will do.

I do not think it would be advisable to shorten their ration of meat. Turkeys require more meat and less carbonatious food than hens, and I am afraid if you increase the corn, before you want to fatten them for the market, you will have liver trouble in the flock. Be very careful how you increase the corn or corn meal.

How Much Grain—I have been feeding three times a day grain morning and night and a mash at noon. I feed a good handful of Kaffir corn, wheat or Indian corn in the scratch pens. I have a mixed flock; I cannot well use the dry mash. How much of the grain should I give if I only fed once a day? I have fifty or sixty hens kept only for eggs and no good way of weighing grain, so please state quantity per hen and not weight.—C. A. B.

Answer—It is a good rule to feed a pint of grain for every dozen hens, the grain to be buried in the scratching pens, so they will have to dig it out. Give all the green food, clover, lawn clippings, alfalfa, lettuce, cabbage, vegetables, that they will eat, and one tablespoonful of green cut bone for each hen, three times a week. You do not mention how you make your mash. Remember that a hen needs animal food, green food and cereals; that is the balanced ration that will give plenty of eggs at all times.

What to Feed and How—Will you kindly tell me what to feed my fowls?

am a stranger in California and cannot make my flock pay for its feed. Four months ago I bought 25 hens and two cockerels (Buff Orpingtons), ten four-months' pullets and twelve Minorcas. The pullets have never layed, the hens only a few eggs. They have new houses and are in an orange grove 100 feet by 65 feet in two pens. I take the Minorcas out of the trees each night. I feed an egg food sold at the supply house here. Grains, alfalfa meal, etc., is in the egg food. The hens have dust baths and I paint the roosts with a lice killer. I get no eggs; one cockerel rattles in his throat. The leading poultryman here has been up and can find no fault. Will you please tell me what and how much, and at what time of day they should be fed? They are high priced fowls and I want to make them lay eggs. The grove is kept cultivated during the summer and everything is new. It seems to be only a question of food and exercise. I get so many different opinions I do not know what to do; some say they are too fat, others not fat enough. How can I make them scratch any more? I would like to feed as cheaply as possible. Where could I get the California Experiment Station Bulletin? —Mrs. L. S.

Answer—Your fowls, especially the Orpingtons, should be laying well. It is as you say, a question of feed and exercise. I find the best results with Orpingtons is to feed grain in the scratching pen in the morning; one small handful scattered in deep straw for each hen. I keep the following mixture in a hopper, or box, before them all the time; also I give them crushed oyster shell, charcoal and granulated bone in a hopper by itself: Mix two quarts of bran, one of corn meal, one of alfalfa meal, one of beef scrap, or of granulated milk. To this I add, on cold days, a tablespoon of ground red peppers, and when they are moulting, half a cup of linseed meal.

If you feed in this way, you cannot fail to have eggs. Besides this, I give the hens lawn clippings, table scraps and refuse vegetables. Hens do much better in this climate when they can have plenty of green food. All the bulletins of the Agricultural Experiment Station can be had by writing to the Director of the Station, University of California, Berkeley, Cal. They are free to residents in this state.

Broken Class for Chickens—Have started in poultry in a small way. Have had very good success so far. However, 'tis somewhat of a trial to get enough gravel or grit for a good sized flock on a small lot. Now, what I want to know is, is pounded glass fit to feed hens? Two of my neighbors have advised its use in the poultry yards, but I am afraid it would act on the chickens the same as it did on foxes we used to poison with it up in the wilds of Wisconsin. —J. G. F.

Answer—Broken glass or broken crockery make a very fair substitute for grit and gravel. It should be broken not smaller than a grain of

wheat and have three sharp edges or corners to each piece. In using glass be sure not to take pointed pieces like slivers, because they may pierce the crop or gizzard. For several years when I could not get grit I used broken crockery for the chickens and I know it does well.

Substitute for Green Food—Will you kindly tell me what would be the quickest and best vegetable for green food I could grow for my poultry? I planted a patch of white clover but it does not seem to grow at all. Is alfalfa meal a good substitute where green food cannot be had?—G. K.

Answer: An alfalfa patch is a good thing to have for poultry, but if you cannot have either clover or alfalfa, plant for the little chickens, lettuce, and for the older ones, kale, swisschard, cabbage, beets, etc. These in the order in which I have mentioned them are the best foods that I know of. You, of course, must judge what will grow best in your section. Alfalfa meal is a very fair substitute for green food, but of course does not come up to the crisp succulent fresh growing greens.

Lack Green Food—I have three pens of White Plymouth Rocks and what bothers me is I only get from four to six eggs from them. They all look fine. I think they are rather fat. As to feed, I give them a small handful of grain in the morning in deep straw, either wheat or barley; about eleven a dry mash—eight quarts bran, four quarts middlings and nearly a quart of beef-scraps; at night I give them the dry grain again. Once in a while a tablespoon of pepper in their mash. They are not troubled with lice or mites, and have grit, oystershell and coal before them all the time; also good clean water. Can you advise me how to feed them so as to get them down to business?—J. B.

Answer: What your hens lack is green food. At least one-third of a hen's food should be green—clover, alfalfa or some succulent vegetables. They cannot do well upon the absolutely dry food you are giving them. Add the green to your present ration and you should get eggs.

Millet Seed—Can you tell me what makes my chickens that are from ten weeks to three months old, droopy? Is millet seed good for little chicks for the first two or three weeks? I mean millet seed alone.—Mrs. P. E. N.

Answer: When chickens are droopy, it is a sign that they may have either lice, worms or indigestion. If you are feeding millet seed, that may account for it. Millet seed is very hard, round and slippery, and passes through the gizzard and intestines without being digested, and I have known of several chickens dying from it. A little used in their food may not hurt them, but an exclusive diet of millet is certain to cause trouble.

Pumpkin Seeds—Are pumpkin seeds injurious to fowls? If so, what would be the symptoms? We feed alfalfa hay, beets, carrots and mangels and our hens are veritable egg machines. I do not think our White Plymouth Rocks can be beaten for size, vigor and egg production.—Mrs. A. E. W.

Answer—I have never found pumpkin seeds injurious to fowls. Mine are very fond of them. If they were injurious they would give the hens indigestion and looseness of bowels. This would be caused by the pumpkin seeds remaining too long in the gizzard and fermenting there.

Ration for Laying Hens—Our chickens are the White Leghorns and they are doing fairly well, at the same time I think they would do better if I knew exactly what to do for them.

Some time ago I killed one that was droopy and found it was too fat. I had been feeding Kaffir corn, so I cut that out and substituted wheat. I see the improvement now.

At present they are laying fine, but the droppings stick to the feathers very bad. I am sure it is not caused by lice, as I dust them with buhach and also paint the roosts with coal oil. I feed wheat in the morning and a mash of bran, onions and red pepper mixed with hot water at night. They also have two lots and a lawn to run on. Do you think they get enough green feed? The range they have is all the green food that I give them.

One of the eggs was covered with a soft shell-like substance in addition to the egg shall and I had to assist a hen pass some of them today. They have plenty of shell and grit, so how can I remedy this?

Will say that I received 202 eggs

from ten pullets in March, and from three sittings of eggs received 80 per cent hatch.—A Reader.

Answer—You are not feeding a rightly balanced ration. An egg-making ration must be composed of animal food. The proportion that gives the best results is about one part animal food to three parts grain or its by-products and two parts green feed. Now, it is impossible for me to know if your hens are getting enough green food because your lawn may be dried up or of tough bluegrass, or it may be succulent green clover. Bluegrass, as the summer goes on, becomes almost worse than no green food at all. And if you have hens confined on two lots I know by experience the lots are soon eaten clean of any succulent grass, and by this time without seeing your premises I feel sure that your fowls are not getting enough green food, and not any meat, so that the ration is unbalanced. The pepper is a strong stimulant and good only in winter. It forces the egg organs for a time, but it will result in a complete breakdown sooner or later.

On close examination you will find that under or on the soiled feathers there are nits, the eggs of the lice. Pull out those feathers, put them into a little can with kerosene in it and burn them, and then dust well with buhach. Give the hens a place well spaded up in the lots so they can wallow in the dust every day. The hens "wallow" should be dug up and turned over at least once a week, so they may have a fresh place to clean themselves in.

The egg shell will be all right if you give them enough green food. If you cannot supply the green food get a sack of alfalfa meal. It will last a long time. Commence by giving the hens only one spoonful of this (for all the hens) in their mash, and every day add only one spoonful extra until you are feeding one-fifth of the mash of the alfalfa at a meal. The hens at first will not like it, but by beginning gradually in the way I have described you will educate them up to it.

Skim Milk—Will you kindly inform me whether skim-milk is a good feed for young pullets or laying hens? Which is best, sweet, clabber or curd? Is there danger of feeding too much curd or skim-milk? Is curd of more value to young stock or to laying hens? I have a bunch of ten-weeks'-old pullets that I am feeding clabber and bran mixed until it makes a crumbly mash. Is it a fattening or muscle or bone making ration? How would it do to feed to laying stock? I give skim-milk to my laying hens in troughs which set in the sun. Will that kill diseased germs or not?—L. E. E.

Answer—Skim-milk is one of the best feeds for chickens or hens at any stage of their lives. It can be fed either sweet, clabber, or curd. By curd, I mean cooked. If you cook it, be careful not to heat it above 100 degrees or it will become tough and indigestible. There is no danger of feeding too much skim-milk or clabber to fowls. The crumbly mash is good feed, but you would succeed just as well by giving them the bran dry and letting them drink or eat the milk as they want it. It is a good bone, muscle and egg-making ration. I give my fowls all the milk I can spare, pouring it into troughs and leaving it till they eat it. The sun does not seem to affect it badly when it is pure milk, but if bran were mixed with it, the sun might make it ferment and then it would disagree with them.

Sprouted Barley—One question in regard to feeding sprouted barley I should like to have answered. Has barley, sprouted, any value other than green food? That is, does it loose its entire value as a grain ration?—L. E. K.

Answer—It depends upon how long the sprouts have grown, and whether there is any grain left. If the sprouts are three or four inches long, there is very little but green food in it. If they have only just sprouted, and about ¼ inch long, there is still considerable grain, but I do not think it is as wholesome as the longer sprouts.

Sorghum Seed—Will you tell me the value of sorghum seed for poultry? Is it fat producing or an egg food and how would it do for turkeys? —C. B. C.

Answer—Sorghum seed, broom corn seed and Egyptian corn have almost the same nutritive value. They can be fed to both chickens and tur-

keys with the same satisfactory results. One year when on the farm I had several tons of broom corn seed which was left where the threshers worked and the fowls had free access to it and the green-growing wheat; they got through the moult early and layed all winter, eggs galore. I never saw better laying and the turkeys did well on it. Professor Jaffa in his most valuable bulletin (Farmer's bulletin 164) on poultry feeding, gives us the nutritive value of broom corn and of sorghum seed as both the same —1:8.4; of Egyptian corn 1:8-6; Sorghum seed is more fattening than wheat and less fattening than corn. If your fowls are on free range and have plenty of green food and animal food or milk, sorghum seed will be an excellent food for them. You should write to the Director Agricultural Experiment Station, University of California, Berkeley, and ask him to send you "Bulletin 164 on Poultry Feeding," then you can see just the right way to balance your ration.

A Tonic and Ration—I want a safety brooder stove or a lamp to heat up a dust box to be used at night. I want to try Mr. Fox's tonic. I have to send off for everything except sulphur, cayenne pepper, charcoal and salt. The others will cost me 85 cents. What is Fenugreek? What will it cost? Please write me out Mr. Fox's tonic again. I wish you would send me balanced ration for Rose Comb Buff Leghorns. How will the following do as a ration and how much will I have to feed to 50 hens a day? Six pounds wheat, 4 pounds rolled barley, 2 pounds linseed meal, 3 pounds shorts, 2 pounds bran, 2 pounds corn meal.—A. V., Chewelah, Wash.

Answer—You can get a safety brooder stove at any good poultry supply house. Dr. Fox's tonic is 10 pounds dried blood or beef meal; 10 pounds ground bone, or bone meal; substitute 5 pounds of linseed meal for the fenugreek; 2 pounds sulphur; 2 pounds powdered charcoal; ½ pound cayenne pepper; ½ pound salt, making 30 pounds in all. The reason I substitute the linseed meal for the fenugreek is because the latter is very expensive here; it costs 35 cents per pound. Your ration is a good one, only you should add five pounds of beef scraps to it, and be sure to give the hens plenty of green food. The green food will make them lay eggs and will keep them healthy. Fifty hens would require about 15 pounds of ration per day.

Will Not Eat Wheat—What is the trouble with my hens? I feed them dry wheat in the evening but they do not eat it, still they have a good appetite. Will you tell me what is the trouble and what kind of food must be fed?—J. S., Phoenix, Ariz.

Answer—Possibly the wheat may be mouldy, or musty. Otherwise your hens cannot be hungry, or they would eat it. Wheat, oats, Kaffir corn, barley, cracked corn, all should be scattered in deep straw or hay, hidden so the hens have to exercise to earn their food. This is the best food, but they also need green food, clover, grass, alfalfa, pumpkins and vegetables. Also animal food, such as meat, fresh-cut bone or beef scraps. They should have charcoal, grit and shell always before them. These are the best foods for chickens.

For a Cold Climate—Last summer I raised 81 Rhode Island Reds and have not lost one of them, nor has one been sick. The climate here is very cold. We have snow until March. This is how I feed: Mash of bran and shorts in the morning; wheat at noon and night. I let them scratch in oat hay for exercise. I give cabbage, carrots, apples and onions raw. Am only getting two eggs now. Should I give my chicks salt and if so, in what way? They have grit, oyster shell, dried bone and dried blood twice a week. I feed them lots of bread; also give them cracklings, and ashes for them to dust in. Will you kindly let me know if I am doing right? If not, where am I wrong?—Mrs. L. B.

Answer—As your climate is very cold, it would be well to mix a little red pepper, a teaspoonful for 12 hens, in their food three times a week. Potatoes boiled and mashed, mixed with oats, a cupful of dried blood added and a little pepper and salt would be very good for them, and will certainly make them lay. It has been found that warming grain such as corn or even wheat, in the oven and giving a feed of that at night during the cold weather will help them to lay. I think you are feeding them all right.

Feeding White Leghorns—I am feeding my White Leghorn hens three times a day, giving them soaked barley in the morning, a good mash at noon and wheat in the evening; every second day they get plenty of green food. Having gone through the starvation process last summer, my eight hundred hens are doing well now, laying between 200 and 300 eggs per day. I have heard much about feeding hens twice a day, so I intend to try this method next spring when eggs are cheap: I will give a good mash in the morning, green food at noon and wheat in the evening. Do you think that would be as good for the hens as three feeds a day? How much mash would be necessary in the morning? I feed a three-gallon pail per hundred hens now, the mash consisting of four and a half pails cracked corn, three-fourths of a pail of wheat bran, two and a half pails of beef-scraps, shells, salt, etc.—J. R. H.

Answer—I congratulate you on your success with your fowls. If you give as you describe, the mash in the morning, green food at noon and wheat at night, it may be called three meals a day, and I think will succeed next spring with Leghorns as well as the feed you are now giving them. I say advisedly with Leghorns, for they are such active, energetic birds that even feeding them a mash in the morning will not make them lazy. With Rocks and Wyandottes, I prefer the mash at night. I like to keep them scratching and exercising in the morning, to keep their egg organs active and vigorous. I think the same amount of mash you are now giving would be all right to feed them in the morning, all the green feed they will eat at noon, and wheat at night, about 12 to 15 pounds of the wheat per hundred fowls at night; you do not say how much wheat you are giving now. If the weather gets cold, damp and chilly, I think it would be a good plan to add more cracked corn to your ration and also more beef scraps. Leghorns require, at least they do better, with a wider ration than Plymouth Rocks; a wider ration is a more fattening ration. The reason for this is their great activity uses up more of the fat, so they require to be fed more of it. A very good plan is, once in a while to weigh your food, allowing each hen about five ounces of solid food, that is without water, per day.

THE EGG QUESTION

Egg-Bound—I have the White Minorcas. Have 15 hens and get from 12 to 14 eggs per day. I have a pullet and an old hen that seem to droop and sit around all day, and sometimes stagger; they had been laying all the time and their combs are still red, but they do not lay now. I feed them bran mash in the morning with alfalfa meal and egg-maker, and once a week chopped onions and red pepper, and at noon we give them green grass, and at night wheat, besides this they get lots of meat scraps from the table; they have oystershell and grit before them all the time. They have not eaten anything since they felt this way, but seem to kind of gasp for breath, and they do not seem to have anything in their craws. Thanking you in advance for a reply, I remain.—Mrs. J. W. S.

Answer: Your hens certainly have been doing very well. Minorcas very often get egg-bound, as their eggs are so large they have difficulty in laying them. This may be the case with yours, and I would advise you to examine them. You might also give them some Epsom salts, half a teaspoonful in a tablespoonful of water. If they are egg-bound, inject a little olive oil and hold the body of the hen in a pan of warm water, as warm as you can bear your hands in; this will relax the parts and enable the egg to pass. If it is indigestion, the Epsom salts will help that. I think your hens may not be getting green food enough.

It Cured Them—How long can eggs be kept for setting and do they require any special treatment? I have a favorite hen and I want to set as many of her eggs as possible, but I do not know how long they will remain fertile, as I have no hen wanting to sit at present. Several of my fowls had a touch of roup and I tried a remedy that you gave (castor oil, camphorated oil, kerosene, turpentine and a few drops of carbolic acid) squirted up her nostrils. I also mixed another remedy that you gave (cayenne pepper, mustard, viengar, lard and flour) and gave it to the fowls, in pills, as you said. I happened to leave it where they could get at it, and found that I need not give it in pills for they were eating it with

relish. I have made the mixture several times since and they seem to be very fond of it. Their combs have become very red and although they are moulting they are laying well. Would you advise allowing them to eat all they want of it? They are entirely well of the roup.—Mrs. H. A. H.

Answer—In reply to your first question it is well to remember that the fresher the eggs you set the stronger will be the chicks. I always set them as fresh as I can get them, and I never sold eggs over a week old for setting. However, I have kept eggs from a favorite hen for three weeks and had a very good hatch. To keep them I always lay the eggs on their side on sawdust or on grain (oats or barley) to keep them from rolling and I turn them every day. By this means the yolk does not adhere to one side, and I have a good hatch. Some advise standing them on the small end, but it does not succeed as well as my way. I am glad your fowls have gotten over the roup. I would not advise you to let them eat their medicine because that remedy is a very powerful stimulant and although excellent for a cold, often curing it in one day, it will prove an irritant if continued too long. It is even now stimulating the egg organs and digestive organs greatly, as is shown by the comb, and I advise you to discontinue it, increasing the animal food; and, as yours are Rhode I. Reds, I would advise adding some oil cake (linseed meal) to the food. This will help to give a fine gloss to the new feathers.

Soft Shelled Eggs—Having read a great deal of your advice I will ask of you a favor. Would you please tell me what can be the reason chickens lay unshelled eggs? They sometimes drop them while on the roost or our among the brush. Mine have been very bad of late; I get as many as three or four a day sometimes from about thirty hens. I should be real thankful to find out what to do for them.—Mrs. L. E. L.

Answer: Soft-shelled eggs are not exactly a diseased condition, but may be a symptom of approaching danger. It is usually due to a lack of shell making material in the food, or to inflammation of the shell forming chamber of the egg duct, which no longer secretes calcareous matter. Over-stimulation of the egg organs by the use of pepper or stimulating egg foods will have this effect. Worms in the intestines may also produce the irritation that will effect the oviduct, and an over-fat condition will increase the tendency to laying soft shelled eggs. This is the common cause of soft-shelled eggs.

Treatment—Provided the cause is an over-fat condition, it can be remedied by giving a ration low in fat producing elements. Give the fowls plenty of shell forming material, such as crushed oyster shells and grit, cut bone and green food; make them work for the grain, which should be wheat in preference to other grains (one heaping teaspoonful to a pint of drinking water) kept before the hens for a day twice a week will help remove the layers of fat. Feed a properly balanced ration and do not try to increase the egg yield by using stimulants that irritate the organs of reproduction.

Poor Layers—Will you please send me a copy of Prof. Jaffa's analysis of foods? I also would like your advice in regard to my flock of hens. I have seventy-five hens and pullets which should have given me some eggs during the past few months, but I did not get one egg through the months of November and December. I have been feeding a mash of bran, shorts, meal alfalfa and meat meal in the morning, and either wheat or cracked corn at night. The hens seem very healthy and hearty in eating, but I notice when I feed the mash in the morning, the droppings look like molasses, but when I leave off the mash and only feed grain, the droppings are natural.

Is meal-falfa as good as Calfalfa? My last sack of meat meal seems rather hard and caked, and I am inclined to attribute the condition of the droppings to this. I am greatly puzzled to know why I get no eggs, while a neighbor who has nothing but common stock and who gives no thought or care to the flock was getting eggs right along. I have thoroughbred Plymouth Rocks and a pen of mixed hens.—Mrs. E. B.

Answer—Prof. Jaffa's analysis of the various foods for poultry in California can be had by applying to the

director of the Agricultural Experiment Station, University of California, Berkeley, Calif. It is called Poultry Feeding, Bulletin 164. I do not keep them for distribution, but you can secure one by sending a postal card directed as above. If your hens got over the moult well, they should have been laying in November and December. The reason they did not lay is either that they did not have green food and animal food sufficient, or that they had not enough lime to make the shells. The cause of the molasses-like droppings is in the meat meal. It may be bad from becoming damp, or from the meat being stale; also it may come from their not having had sufficient meat in the last two months. When first they get the meat, they may over-eat of it and that has an effect on their bowels. Any change must be gradually made. A hen should have about three ounces of grains, or their by-products, two ounces of green food and from half ounce to an ounce of animal food. This, with proper exercise and cleanliness, and plenty of clean water, will insure egg production.

Calfalfa and meal-falfa are one and the same. They are alfalfa hay well dried and ground, and are excellent for poultry. As you have Plymouth Rocks, I would advise you to cut out the corn. It is fattening. You do not mention the quantities you are giving of each ingredient so it is very difficult for me to guess where it lacks.

Over Fat Hens—I have about two dozen Buff Orpington hens and have had no eggs for four months. They appear as healthy as can be. For some time I fed them wheat twice a day and the table scraps. I began to think I was not feeding the proper foods; then I got bran and an egg maker and also bought cabbage for them and still no eggs. They have lots of exercise and gravel and are so fat you cannot eat them. Please tell me what to do to reduce the fat. The past two weeks I have been giving them just the scraps from the table. Tell me, is that the proper method to reduce fat?—Mrs. A. C. S.

Answer—You hens are so fat that they cannot lay. The whole inside of them is filled full of fat so the eggs cannot pass down the egg duct. The best plan would be to kill and eat, or sell the fowls, because they will not make satisfactory layers after being so fat.

However, if you wish to keep them, your only plan will be not to give any grain, or any table scraps until they are reduced in fat; give only green alfalfa or lawn clippings, for two weeks, then commence and feed half an ounce of meat per hen per day and lawn clippings; no grain or bread, and in about a month they may begin to lay.

Hens Stopped Laying—You answer so many questions that I venture to ask a few. I have some hens, seven of them, ten months old that were laying; recently I bought a young rooster; the hens stopped laying after he came; sometimes one or two of them lay. I feed wheat, table-scraps and cut alfalfa for them every day; I used to let them out but cannot do so now. Is a good handful of wheat enough for a hen at each feed three times a day? I keep dry bran by them all the time; also crushed oyster shell.—Mrs. M. W.

Answer—The reasons your hens have stopped laying is because you have stopped letting them out. It has nothing to do with the rooster. When they ran out the exercise kept the egg-making organs active. Feed according to the directions I have already given and in a few days you will have eggs and more of them.

Good Laying Pullets—Could you tell me if a pullet, hatched from a hundred and fifty egg laying strain that begins to lay at exactly the same age as the pullet from a two hundred and twenty egg strain hen, is just as good a layer as the pullet from the 220 egg strain hen? Would the pullet from the 150 egg strain hen be just as good as the 220 egg strain pullet if she moulted as good? Why is it that a hundred hens on the same space of ground will not do as well as fifty hens when their yards and houses are kept clean?—R. T. S.

Answer—"You can't sometimes always tell!" Usually the pullet that commences laying the earliest is the one that lays the most eggs, but very much depends upon the care and feeding. If you have two pullets as good as you think, keep track of them both and let me hear from you

at the end of a year. It would be quite interesting to watch them and find out which is the better layer. It has been proved beyond a doubt that when a small number of hens are kept together, they lay more eggs pro-rata than when large numbers are kept together. There are many reasons given for this but I do not know that any have been proven to be correct. It is, however, a well known fact.

Blood Spot on Yolk—I have 150 Brown Leghorn pullets just starting to lay, and I supply a few customers with eggs and they have been complaining of finding a little blood spot on the yolk. I have plenty of nest room so they are not crowded. I have been picking 70 to 80 eggs a day. They have abundance of green feed. I feed soft feed in the morning, wheat at mid-day, corn at evening, so if you will please let me know what the cause of this is I will be very much obliged, because my customers are getting dissatisfied.—W. W. M.

Answer—The small blood clot you describe results from a slight hemorrhage which has generally occurred in the upper two-thirds of the oviduct. Such hemorrhages are the result of great functional activity and congestion of the blood vessels. They are excited by any of the causes which lead to congestion and inflammation and are to be counteracted by green feed and less animal food and by the suppression of red-pepper or any stimulants. Give a little Epsom salts in the water and add about twice the amount of salt you are giving to the mash in the morning, leaving off the red-pepper.

Best Layers—Will you kindly tell me what breed of chickens you consider best for laying? I have fifty chickens—mixed varieties—from one to two years old. I am feeding them about three quarts of mixed grain per day. They have layed about seven eggs per day for the past month, during which time they have been moulting. I also give my chickens beef-scraps, oyster-shells, green stuff and plenty of fresh water. Can you advise me what I should do to make them lay more?—Mrs. W. J. N.

Answer: I consider the Standard bred chickens the best layers, as they have nearly all of them been bred-to-lay. Your hens, you say, are moulting. You have to give them sufficient food to make the new feathers, as well as to lay eggs, and you are not feeding your hens enough. I have thirteen hens, moulting, and they are laying from six to eight·eggs every day, and I feed them nearly as much as you give to your fifty hens. It pays me to feed well because I get plenty of eggs. If you want your hens to lay well, keep beef scraps before them all the time, and feed them more liberally of grain, and all the green food, clover, alfalfa or vegetable, that they will eat.

Largest White Eggs—I am starting or trying to start a poultry ranch and would like to ask you a question recently asked by some one else but in a little different way. Which of the good laying breeds lay the largest white eggs? My aim is for good city trade.—E. A. M.

Answer: The Black Minorcas have the reputation of laying the largest white eggs. The White Leghorns are their close competitors. It very much depends upon the strain or family. For instance, one set of fowls may have been selected for beauty of feather and form and their owners may not have chosen those that layed the largest eggs, whilst some have carefully chosen the largest egg-layers, and bred from those, not caring for exhibition birds, and again a third party might have united these two qualities and have both prize winners and the best of layers. It depends upon the ability of the breeder and also upon his object.

Black Minorcas do admirably in the climate of Southern California. I do not know how they would grow in a damper, colder climate. You would have to inquire of people who have had experience in that kind of a climate.

Egg-Bound—Will you please tell me how to treat one of my hens? She gets on the nest every day to lay, but she does not lay. She is not a young hen and does not eat her eggs. I go to the nest just as soon as she gets off. She has not layed for almost two weeks.—J. E. D.

Answer: Your hen may be egg-bound, or she may be just commencing to moult, which I think is probably the case, especially if she has been a good layer. She may be a very poor layer

if she is from debilitated or inbred ancestors.

Sudden Death—Lately I have had three hens die suddenly, and apparently without cause; my neighbors have also lost several. Perhaps you can enlighten us and suggest a remedy. The hens were laying, combs red and large, crops full of wheat, etc., but die on the nest over night. I held a post mortem examination and could find nothing radically wrong. Each had well formed eggs and many of them. They roost high in the open air; run out nights and mornings on alfalfa. I feed wheat mostly, and once every other day hot bran mash with a spoonful of egg maker. Have had over 40 dozen eggs without interruption since January 1st from twelve pullets—Minorcas—of my own raising. This is the first death I have ever had except of the little chicks. Pens are clean, no lice or mites. Have studied closely and can't "savy." Perhaps you can. The heart of the first one seemed the only cause for death, as it had a large inforct, probably fatty degeneration; the other was normal.—Dr. J. A. B.

Answer: I think as your hens died on the nest, that they had some difficulty in laying, and were probably egg-bound. The Minorcas laying a large egg, are frequently subject to this trouble, more so in fact than the other breeds which lay smaller sized eggs. Straining in laying frequently is the cause of a blood vessel breaking in the head, which, of course, results in apoplexy. Minorcas rarely suffer from an over-fat condition, as they are a very active breed.

Way to Get Hens to Lay—What months should Black Minorca and White Leghorn chickens be hatched in order to get them to lay from August 15th on? How many little chickens are necessary under fair conditions to raise 500 hens? What kind of an our-door brooder is necessary to raise say 500 chicks at a time? Which is the cheapest for a person not experienced in incubators—to buy chicks just hatched or to hatch them? What feed do you recommend for little chicks? Do you think it possible to raise chicks on adobe soil? Summing up the main motive of this letter, please advise cheapest and best way I can get 500 hens ready for laying by August 15th.—S. D. W.

Answer—In order to get the Mediterranean class of fowls to lay by the 15th of August, I would advise that you hatch one third in February, one third in March and one third in April. In this way you will have a succession of layers just when your old hens are moulting. The eggs usually hatch half pullets and half cockerels, so in order to have 500 pullets, you would have to hatch at least 1000. You would have, of course, to make allowance for losses in raising, so you had better count on hatching 1200. To raise 500 chicks, you should have ten out-door brooders. Any of the standard makes are good. I would advise a person not experienced with incubators to buy chicks just hatched, if they are property hatched. I recommend dry chick feed for the little chicks. Adobe soil is bad for chickens, but by getting a load of gravel for the chicken yard you can raise them.

Egg-Eating Hens—Would you kindly tell me how to treat egg-eating hens? What will cure them?—Mrs. R. E. G.

Answer—The best way is to cut the head off the offender and eat her, for she is certain to be fat. The information you ask for is as follows: Mr. Morse (a chicken expert) gives five remedies for the bad habit of egg eating. First: Fit up an arrangement whereby the eggs as soon as layed, slide down and out of sight, into a sort of false bottom under the nest. The hens will not eat them because they cannot get them. Second: Have a lot of china eggs lying about promiscuous like on the floor.

Trying to eat such eggs is likely to discourage egg-eating. Third: Fix up a hollow egg with aloes. One bite is enough. Consult the corner druggist as to how to make the mess. Fourth: Have grit and crushed oyster shells about in abundance in self-feeding boxes. Fifth: Do not stuff your hens full of mash in the morning and let them sit around all day, like "Father" in the song "Everybody Works But Father" but feed them grain in litter and make them hustle all day. This keeps them out of mischief. Mr. Morse's advice may be good, but I recommend using trap nests by which means you will easily discover the guilty hen and if she is not too valuable, the verdict should be decapitation. Keep oyster shells, grit and charcoal before your hens and there will be very little egg-eating for it is a vice which always commences with weak or soft egg shells.

Novel Nests—Do you know the name of the maker of a nest with an opening in the bottom so that the eggs will drop through into a box below to prevent the hens from eating the eggs?

Answer—I have seen the mention of such nests but have never in all the many poultry ranches I visited seen such nests in use. You might try darkened nests. They are simply a curtain of burlap hung in front of the nest with a split up the middle. When the hen has layed and stepped off the nest the curtain closes behind her and she can not see the egg to eat it. This has been found successful.

THE MOULTING SEASON

Forcing the Moult—Since reading your article on forcing the moult I have decided to try it, and would like to know if it will be all right to feed alfalfa only for the three weeks, as that is the only green food we have. How long to feed it, and should it be before them all the time?

We have been feeding our chickens cracked wheat and kaffir corn, but they are very bony and many of them have white feathers.

These chickens are Black Minorcas hatched in March and we think we have not been feeding them right. Charcoal, grit and oyster-shell are provided and cool water fresh three times a day, green alfalfa every noon and green bone twice a week.

Our hens are fed by the dry hopper method and we give them whole wheat in the morning. Do they need anything at night? What would be a good way to start them with the grain feed after I have forced the moult?

We have sixteen Black Minorca hens which are almost bare of feathers on their backs and breasts, which brought us in $7.00 worth of eggs in August. Do you think they are good layers?—Mrs. E. E. C.

Answer—I am afraid you have not quite got the idea about "forcing the moult." The reason for putting the hens on short rations is to stay the laying and to draw the nourishment from the old feathers, making them dry and lifeless so they will easily fall out. When the feathers get so they will fall out quickly, then—then only —we feed them a rich food, but supposing the feathers have fallen out, we do not then want to starve them, or the feathers will not come in, or if they do they will not be bright and lustrous. When the feathers are forming or are coming in the hens should have a generous ration of feather forming food, corn or corn meal, linseed meal and animal food. This will hasten the forming of the feathers and will "force" the moult.

Now, you say that your hens are nearly bare of feathers, therefore I conclude they do not need to be made to shed their feathers, but are in the condition that requires plenty of good, rich feather making food. And the sooner you can get the new feathers on to them the sooner you will have winter eggs. You certainly have good layers if you had such good returns in August.

About the younger chicks, if you feed them more animal food than you seem to be giving them they will develop better and their feathers will come in black, when they are mature. I think with you that you have not fed them the right ration or they would be large enough to be laying before the end of this month.

The Proper Month—Is not the month of July about time to force the moult? I want my hens to be laying this winter when eggs are fifty cents per dozen, so I want them to moult now. Also do you recommend constantly keeping copperas in the drinking water? Do you think the scraps from the table, along with wheat, a good food?—A Beginner.

Answer—July is too early to force the moult; August is early enough. I do not recommend keeping copperas in the water, because healthy hens do not require it, and I am averse to giving drugs or stimulants unless for sickness. The scraps from the table are excellent for chickens, especially the meat and vegetable scraps.

To Hasten Moult—I would like to ask your advice regarding my chickens. They have been moulting for the last ten weeks and of course, did not lay any eggs. They are Buff Orpingtons, and are very healthy. I feed them twice a day, bran and sour milk and in the evening before they roost, wheat and also Swiss chard and oyster shells. Could you please advise me what to give them to get them through the moult a little quicker?—V. V.

Answer—Give your hens a little corn meal (about twenty per cent added to the bran and milk) and some linseed meal (about five per cent); also either add some beef scrap to the mash or give them beef scrap with the oyster shell in a box where they can eat as much as they want. They are not getting enough nitrogenous food to make the feathers come quickly. With this addition to the food, they will be soon through the moult and begin to lay.

Stopped Laying—What can I do for my hens? They have stopped laying for the past week, and am at a loss to account for the same. They are a healthy stock with bright, red combs. Have about 30 laying hens and only one Buff Orpington rooster among the lot? Do you think that sufficient? I feed them bran mixed with cayenne pepper—a very little— also an egg maker. Any information you can give me will be greatly appreciated—N. B.

Answer—Your fowls are evidently beginning to moult. I notice that a great many fowls are beginning early this year (July, 1907). I cannot account for it unless it may be the spots on the sun! Help them through the moult and they will begin to lay again.

One Orpington male should only be mated to ten hens of his breed to produce fertile eggs.

A Little Late—I am just a beginner and have only a few chickens. Only a week ago I bought my last three and yesterday I discovered that one of these had shed all of the feathers from off her neck, leaving only pin feathers sticking out. Under her wings are great bald places and her feathers can be pulled out easily anywhere you touch them. Her flesh is dark blue and her bowels seem to be loose, otherwise she is apparently well, scratches and eats just like the other chickens. All over her head the feathers are off and there are a few small white spots on her comb and one quite black one.

My other hens are very healthy, get all the green clover and other feed they can eat, and lay well. I do not think the chickens got much green food before I bought them.

Do you think Blue Andalusians a very profitable breed compared to others?—Mrs. J. B. D.

Answer—Your hen is old, consequently is having a very late moult. This is being helped on by your good care and change of food. The blue color of her skin is often there when moulting late. The spots on her comb are chicken pox. Put carbolated vaseline on those. Keep her by herself until she is feathered out and quite well again. Give her plenty of green food and plenty of animal food. Andalusians are a very good fowl and compare well with any, but I would not advise you to keep more than one breed at a time.

Nature's Way—Your article on moulting interested me very much. Will chickens hatched in March, April and May moult this fall? If so, should the pullets be treated same as the hens? I mean by the fasting method. Will they moult as early if they have not matured rapidly?—K. H A.

Answer—Pullets hatched in March, April and May will not moult this summer. They will take on some more feathers to help keep warm during the winter months, but will not shed feathers except a few of the earliest. Do not make them fast. A pullet should be well fed from shell to shelling. They will not moult early if they have not matured early.

HATCHING WITH INCUBATOR AND HEN

Poor Hatches—We have been running our incubator since February and our hatches have been quite poor; our hens are two years old and so are our roosters. The hens are fed regularly, and have a large run with plenty of alfalfa; a clean airy coop.

The chicks when hatched are strong and vigorous. We have some six weeks old and we have not lost one, but when they are hatching many die in their shells. Out of 450 eggs 77 tested out not fertile or dead germs, and out of 373 remaining eggs only 182 hatched. We are hatching White Leghorns. Can you tell us what to do, or what the matter is? We have been following your advice in many things.

Do you think that slamming of doors or jarring is bad for incubators when hatching?—Mrs. M. F. De W.

Answer— I think the fault in your incubator is that it has not sufficient ventilation. An insufficiency of oxygen will cause poor hatches such as you describe. With the care you give your fowls and their being two years old, the fault does not lie in the parent bird or their eggs, therefore it certainly comes from a faulty incubator. In the future air the eggs three times a day; fan out the stale air of the incubator each time you air the eggs, and if you find they are drying out too much sprinkle them, after the first week, twice a week with warm water. Slamming the doors or jarring the incubator during incubation is not advisable, but on the day of hatching it would not injure them.

Infertility—Will you kindly tell me what to do to make eggs more fertile? I have a fine pen of Columbian Wyandottes, eight pullets mated with a cock two years old. They are fed on dry mash of bran, ground barley corn meal, alfalfa meal and beef scrap with plenty of grit, shell, charcoal and ground bone before them all the time, and are running in a corral of grass and clover; they have plenty of fresh water and the hens lay well. What chicks I do get are strong and healthy; out of fifteen eggs only two were fertile.

I have another pen, four hens two years old mated with a cockerel one year old. Fed the same in every way; their shells are smooth but full of clear spots. What shall I feed to make shells better?—Mrs. E. H. G.

Answer—The usual requirements missing, from the food when eggs are infertile are green food and animal food, therefore I would advise you to feed more green food, more animal food and a great deal less barley and corn meal. Wyandottes are apt to get too fat to have good fertility unless they have plenty of exercise. From your account, I think neither pen has sufficient exercise and the four old hens require more lime. Mix some fresh quick lime in water to the consistency of pancake batter; let it stand 24 hours, then pour out a cake of it on the ground. It will soon dry, and by crumbling a little of it every day, the hens will pick it up. Add a teaspoonful of baking soda to a quart of their drinking water and keep this before them for a week. By this means I think your eggshells will improve.

Airing Eggs in Incubator—You have stated that you aired your eggs about one hour daily. Would that have a tendency to make your hatch come off late, or did you run the machine higher to offset the cooling? Did you start in from the first week to air that length of time, or was it gradual? If I aired them longer without chilling, could I get them out in time, or does airing them make them late? The chicks that came out were very wet; some of them stuck in the shell; the stuff drying down and glueing them in.—Mrs. N. A. R.

Answer—After the eggs have been in the incubator 48 hours, I commence airing them about five minutes twice a day, gradually increasing the time two minutes each time. By the third week I am airing them 20 minutes twice a day, or if the incubator is a hot-water machine, I air them three times a day in a room that is not lower than 70 to 75 degrees, because I do not want to chill the eggs. If they are too much chilled or cooled off, they are apt to be weakly, the hatch retarded and the chickens have difficulty in coming out of the shell, such as you describe. Evidently you have either cooled the eggs too much or you have run the incubator at too low a temperature. We want to give the eggs as much oxygen (fresh air) as possible without chilling them.

Cripples—Some of my incubator chickens are almost cripples when they are taken from the incubator. Some have crippled, crooked and crumpled up toes, others have one leg too short, or turned out the wrong way, and some of them are not able to stand up—they hold their head back so far that they fall backward. —A. H. S.

Answer—The cause of cripples invariably is irregularity of temperature in the incubator. Your incubator has been too hot at some period, probably the last week; this causes cripples. Those that hold their heads back do so from the eggs not having been turned sufficiently during incubation.

As you do not mention the name of the incubator, I cannot tell you just where the lack is. It may be poor oil; it may be it is run in a draught and it may lack ventilation.

Lack Oxygen—I took 200 thrifty chicks from the incubator about eight weeks ago. They did very well for about two weeks when they began to die and today I have 50 left, and these look too scrubby to be worth raising. I have given them extra attention and the best feed. They get pale around the head, grow weak and are skin and bone when they die. I think they have consumption. The brooder is a tight box and no ventilation, except the lid has a round hole about as large as a teacup, and the little entrance window about six inches square. An iron pipe running through is the heating arrangement. Inside the box to fit close over the pipe, is a cap of wood with flannel curtains dropping to the floor under which the chicks hover. Don't you think this is too close a place? The outside box is only 6 inches deep, then they hover inside; this only gives 4 inches space for the chicks. Please tell me if you think the lid to brooder would be better of wire or where do you think the trouble is? Also tell me how granulated milk is prepared. We have late-ly begun feeding to everything in the poultry yard beef scraps, bone meal and linseed meal in what we think proper proportions once a day. Should chicks only eight weeks old be fed this ration the same as hens? What causes eggs to be ridgy and uneven? Can one feed to produce larger eggs? Our hens are large but lay small eggs.—Mrs. J. B. S.

Answer—I think that the lack of oxygen in your brooder is the only difficulty with your chicks. Still I am very much afraid that tuberculosis may have got in, and infected the brooder. If possible, move your chicks into a weaning house, open entirely on one side (or only closed with chicken wire). Make a little frame of gunny-sacking or out of a piece of blanket that they can go under. This will rest upon their backs to keep them warm. Give them no other heat. At this season of the year (August) eight weeks old chicks should have no heat whatever at night. I think you are keeping your chickens too warm, without enough fresh air and possibly they may have mites or lice. Air their sleeping place well; put the hover out into the sunshine every day. This will kill the germs of tuberculosis better than anything.

Granulated milk is made at Binghamton, N. Y. I do not know the process.

Chicks eight weeks old can have the beef-scraps, bone meal and linseed meal in the same proportions as hens.

Uneven eggs are caused either from defect in the oviduct or from an insufficiency of lime or hurried laying.

Some strains of hens lay small eggs and over-fat hens will lay small eggs. More protein added to their food will often increase the size of the eggs. By choosing the large eggs for hatching, you can increase the size of the eggs in the next generation.

Setting Hens—Can you tell me what is the matter with my chickens? They seem good and healthy until they start to set, then they invariably develop a severe case of diarrhoea, which causes them to leave their eggs after a few days. I have now a hen that wants to set, and have just received a setting of thoroughbred eggs, but today I noticed the same trouble as with the others, except that she seems to be a great deal worse, for her droppings are of a bloody nature. Can it be from too much bluestone in their water or because of too much egg-food? I feed them a mixed food from the feed yard, consisting of corn, wheat, Kaffir corn, beef scraps, hone, charcoal, oyster shell, barley and some other grains I cannot classify. They get this twice a day together with all the table scrap and all the grass they can eat. They also have plenty of exercise. Is there anything I can do for this particular hen? Shall I try to set her or get some other hen for the eggs? Still another question, what causes a milky, watery substance in the whites of the eggs; it runs out after the eggs have been cooked?—G. W. Y.

Answer—It is the bluestone in the water that thoroughly disagrees with or poisons the setting hens. Feed a setting hen only grains, wheat and corn mixed, and give her fresh water to drink without any medicine in it. You should not be giving your hens bluestone at this season of the year at all. They do not need it, and it will injure the fertility of the eggs and make the chicks hatching out weakly. Do not set the hen you mentioned, as in all probability she will leave the eggs. All setting hens should be in perfect health and entirely free from lice or mites. You had better get another hen for those eggs.

The milkiness in the whites of your eggs is an indication that they are perfectly fresh, that is, new layed, and is a great recommendation for the quality of your eggs.

Chicks Dying in Shell—A large per cent of my chicks fully developed die the day they are due to hatch, even after pipping the shell. They seem to dry in the shell.—Mrs. D. D.

Answer—Float the eggs in warm water. That will help the chicks to break through the shell better than anything I know of. Next time try sprinkling the eggs after the eighth day twice a week with warm water. I think you will find it is what is needed in your dry climate, and is likely to help matters.

Fooling the Hen—Is it possible to fool a sitting hen into caring for some incubator chickens when she has not hatched them herself.—Mrs. C. R.

Answer—If your hen has been sitting for a week or ten days, she will "take to" the chicks as well as though she had hatched them herself; especially if she is a Plymouth Rock or Buff Orpington. Those two breeds have a greater affection for chickens than some of the others. Be sure that the hen is entirely clear of lice, and if she is a large hen, put from 15 to 18 under her at night; a smaller hen should have from 12 to 15, not more if you expect the chickens to do well. I have trained capons to act as mothers; they do even better than the hens.

Thermometer—Will you kindly tell me where I could get tested thermometer for incubator; also where I could have one tested which I already have?—H. H. C.

Answer—At any good drug store you can have your thermometer tested. If you want to buy a new one, go to the agent selling your make of incubator. Take the new one also to the druggist and have him test it thoroughly, because the thermometers as they are seasoned sometimes vary degrees, and even a new one cannot be trusted.

Helping Them Hatch—I find my White Plymouth Rock eggs are very slow about hatching and some I know would die in the shell if I had not dropped a few drops of lukewarm water on their heads, as it seemed they would get about half out and then the white skin would dry on their heads and hold them fast. After having two die in the shell I found they would free themselves if a few drops of warm water were sprinkled on them. I kept moisture in the pans all three days and part of the fourth and they are still slowly hatching. This is the twenty-third day. Do you think I should keep the moisture pan full for a week—I mean the last week of incubation? Please send me an idea on chick feed as I can not get good clean chick feed here.—Mrs. P. W. B.

Answer—If you had only mentioned the name of the incubator you are using I could have better diagnosed your case. As it is, all I can say to you is to follow the rules and directions they give you as closely as possible. With some machines it is very advisable to sprinkle the eggs twice a week after the twelfth day with warm water; this seems to make the shells more brittle and prevents the inner lining skin from toughening. I have found this better than keeping much moisture in the machine. The moisture in the machine seems to make the chick grow but does not make the shell brittle. Your Plymouth Rock eggs should hatch promptly on the 21st day. The delayed incubation indicates that part of the time the temperature has been too low. Are you sure that your thermometer is perfectly correct; have you had it tested? On the efficiency of the thermometer much depends. Many thermometers that are accurate at first become, through the use of unseasoned glass in their manufacture, absolutely incorrect after a few months' use. Others are really only within two to four degrees of being correct, therefore be sure you have your thermometer tested. About the chicken feed, write to the Experiment Station, University of California, Berkeley, for bulletin 164 on poultry feeding. This gives you the lists of foods available in your part of the country, with the proper proportions for mixing them.

Eggs for Hatching—Will you kindly tell me what is the matter with my eggs? They will not hatch well. Our hens are Brown Leghorns and Rhode Island Reds. I only got fifteen chickens in my last batch. When we broke the eggs after we know they will not hatch we find the chicks dead, but fully formed and just ready to hatch. Perhaps the shells are too hard. Will you please tell me what to do to make a softer shell? Feed according to your directions.

Is it necessary to put moisture in the incubator? Does it hurt the eggs to sprinkle them with warm water if we think the shells are too hard? I will be very thankful if you will answer this, as I want to know before I commence to save eggs for next incubator lot. I do not keep them over two weeks and keep them in a cool, dark place, turning them every day. —Mrs. G. A. M.

Answer—I wish I could tell you for certain what causes chickens to die in the shell. I have my theories about it, and I believe it comes from the eggs not being aired and cooled sufficiently. Cooling them and then warming them up again seems to make the shells more brittle, and this is the same under hens. If I notice that a hen is setting too closely ?

take her off twice a day to cool the eggs. With an incubator I would air them and turn them three times a day, and either sprinkle them three times during the last ten days or float them in warm water two days before the hatch is due. Float them from three to five minutes and then put them back into the tray while they are wet. I do not believe in putting moisture into the incubator unless the directions call for it.

Incubator Chicks Dying Off—We have started in with the R. I. Reds, and have been fairly successful until our last hatch. Out of 65 eggs 44 came out. Last Saturday they commenced dying off, just fell seemingly from weakness and died soon after. We have fed them chick feed, bran, Indian meal, cayenne pepper, beef scraps, twice per day, and a little germazone in water occasionally.— C. R. H.

Answer—From your description, I am afraid that the chickens have either been chilled or may have been over-heated. Either one of these conditions will cause the symptoms you describe. All you can do now is to give them rice boiled in milk, adding a tablespoonful of ground cinnamon to each pint. Give them also chopped lettuce and onions. Do not give any cornmeal or beef scraps. When chicks have been over-heated either in the incubator or brooder, it so weakens their bowels that they cannot digest their food and they die of starvation.

Poor Hatching—I should like very much if you can give me some information about my hatching eggs in an incubator. I bought a new incubator this spring. I have set it twice and had the same results both times. The chicks form fully and then most of them die in the shell. As the same eggs do fine when put under a hen, I think it must be that I make some mistake in my treatment of the incubator. I have as nearly as possible followed the instructions that came with it. If you can give me any assistance it will be appreciated very much.—Mrs. W. D. W.

Answer—Your incubator is a good one. Its fault, for they all have some little fault, is that the ventilation is insufficient. Take the eggs out and air them after the first week three times a day. This will counteract the lack of ventilation. This cooling and then heating up again of the eggs makes the shell more brittle, so that the chick is able to break its way out much more easily. Another thing I found in using that incubator is that by taking the middle eggs out of the row, one in each hand, and putting them at the end of the row, and then pushing the others along into the vacant places, I got a ten per cent better hatch. I got the idea from Egypt. Of course, you must be sure the machine stands level and that the thermometer is correct.

———

Trouble with Incubators—I want to ask your advice about our incubator. We bought it new in January. Out of 200 fertile eggs we got 75 chickens and all but nine died before they were 10 days old. We thought it was the fault of the brooder. There were many cripples among them, but they all died of bowel trouble. On April 30th we hatched 117 out of 150 fertile eggs and gave the chicks to old hens, as we had laid our previous trouble to the brooder. But now the last are going the same way. Chicks hatched under hens at the same time are healthy and strong. We have only lost one so far. We feed prepared chick feed and take the best of care of the chicks. The incubator runs perfectly, always 103, until the chicks begin to work out of the shell, when it runs up to 104 and 105. We have set the incubator again. It will match May 29th. We do not intend to give up.—W. S. R.

Answer—The trouble is in the hatching. At some time or other the heat has been too great. This is shown by there being cripples. I know it, because I have had the same experience several times myself. Once a hat was thrown on the machine; just touched the regulator; was only on for half a day. Another time a newspaper did the same thing. My big cat slept on the incubator another

night and lost me the hatch. Each of the times I worked with the little chicks, giving them everything I could think of, but without saving them. Now, I think there is a possibility that your incubator does not stand level and that, therefore, one side or corner of the machine is a very little higher than the other. That side or corner would be hotter than the other side without it affecting the thermometer and would cause all or most of the trouble. Again, are you sure the thermometer is correct? Borrow the doctor's clinical thermometer. This is what I did and put them both into a bucket containg about two quarts of water at 103 degrees and compared the two. You do not mention if the hatch came out on time. I feel sure that the eggs have been overheated, or part of them have, and in this way the bowels of the chickens have been weakened, the yolk of the egg has not been digested and they have dwindled and died, or bowel trouble has come on from the indigested yolk purifying inside of them. I have made so many post mortem examinations that I feel sure of what I am telling you. Examine your incubator with a spirit level to see that it is level. Test your thermometer and then try again, at the same time setting one or two hens, and as incubation proceeds examine the eggs, comparing them. I think you will find that the eggs under the hen dry out less quickly than those in the incubator. However, if this is not the case, if your incubator eggs dry out too quickly (the air space being larger than that under the hens), you will have to regulate this by the ventilators of the incubator. Keep them closed. As yours is a hot-air incubator there is no need of fanning out the stale air. The fault, if any, with your incubator is too rapid a circulation of air, thereby drying the eggs out too soon. I think you had better run it half a degree cooler than you have been doing. I say this because the cripples and bowel troubles denote too high a temperature. I hope these hints may help you. Let me hear from you again if you have any more trouble.

POULTRY HOUSES

Mushroom Houses—Will you kindly publish a plan of a Mushroom house? I expect soon to have some choice Minorcas and wish to put them in the most suitable houses.—L. J. H.

Answer—In this book is a plan of the mushroom house. I have used them and found them quite satisfactory, except for a few defects. These were that when they were made like the cut, ten inches from the ground, the chickens would get down in the morning off their perches and were in a draught until the attendant left them out; also it was difficult to reach the chickens to handle them at night, so that I had to make a door at one end, or side. I found the open front houses more satisfactory. These houses are made perfectly tight on three sides and the fourth side is either open entirely or closed partly either with burlap or wood. See pictures of these elsewhere in this book, which are used on many ranches hereabouts.

Mushroom Houses—1. In building so-called mushroom chicken houses, shed roof style, how high should the lower corner of the roof be above the ground?

2. How high should the lower wall be?

3. How high above the lower edge of building should the roosting poles be placed?

4. How far from the ground should the lower edge of the house be?

5. How do they manage in Petaluma to get good winter layers? Does each poultry man raise his own eggs for hatching?—C. W.

Answer—1. From two to three feet.

2. The lower wall should not be less than two feet in height.

3. The roosting poles should be from six to twelve inches above the bottom of the wall, enough to keep the hens out of any draught.

4. I prefer the lower edge of the house to be only four inches from the ground as that gives ample room for ventilation, but in this way you have to make a trap door for the hens to get in and out; to avoid this some people make the house to stand higher from the ground, or about ten inches, placing the roosting poles also higher, or about two feet from the ground. With the lighter breeds such as Leghorns this does all right, but heavier birds are apt to bruise their feet if they have to fly down from that height.

5. In Petaluma the poultrymen nearly all raise their own chickens, but some buy the young chickens from some excellent hatcheries they have there, and where they get them well hatched at one day old. They get the winter layers by having early hatched pullets and feeding them right.

More About Poultry Houses—It is with great pleasure I take advantage of your invitation to write you. I have been taking the Orange Judd Farmer for several years. It is quoted as a supposed authority. I have also several Standard poultry books. But for solid facts, and nut-shell information, the poultry department of the Live Stock Tribune gets away with the cake. Now the facts in chicken-house building are just what I am after at the present time.

Reliable and practical information as to the successful kind of chicken houses. I may say I am considered a first-class carpenter, so the labor will not hurt me. I do not wish to build anything in the nature of a luxury, but practical, up-to-date convenience, and a nice size to keep chickens healthy. One of the most important questions with me is ventilation without causing disease. Have several types of coops in mind. First, open on one side and enclosed on the other three. Second, open all around below the roosts (which class I think would need ventilation at the ridge, and I am afraid would create injurious draught from below). Third, open above and below roosts, and tight boarded all around upon line of roosts.

If you could suggest something better than any of the above, I should be more than pleased to hear of it. From your articles I have the faith in you that moves hills and even mountains.—A. J. R.

Answer— You have decidedly formed the right idea of ventilation; there must be pure air but no draughts in a hennery. Your first plan is the best. You must have the open side turned away from the night breeze. There are pictures of houses of this description in this book. This is called the open-air house.

Your second is the mushroom house, so called. It also is good, but must have no opening, or ventilation at the ridge, as that would cause an injurious draught. I have tried both of these plans, and find them both admirable. In fact, fowls of healthy parentage will never be sick in houses of either plan if they have sufficient green food and pure water. The air in these houses passes over the droppings and carries off the effluvia, so the house never smells close, and in the mushroom house, the heat from the bodies of the fowls is conserved in the top of the house and they do not feel the chill of the night air, which draws upon their vitality, consequently upon the egg production. These are the most sanitary arrangements for henneries in this climate I have tried. The third plan you mention, would not be satisfactory. It would be too draughty.

Housing Chicks—Do you think it advisable and safe to have four months old chicks in a house facing towards the north with boards for half and burlap for the other half of the front? If not, would canvas be better? Is canvas waterproof enough for a chicken house?—C. W. S.

Answer—I think it quite safe to put four months old chicks in a house facing the north such as you describe. Give the chicks plenty of fresh air but no draughts. A crack or a knot hole will give colds, bronchitis and roup. Canvas is water-proof enough for a chicken house and makes a good chicken house. You can easily find if it leaks, and mend the hole, or you can even paint it with oil.

Whitewash—Is there a good cheap way to make whitewash that will stay inside and out of poultry houses and stand rain and sun without scaling off so soon?—"B."

Answer—Here is a recipe for whitewash which is unrivaled. It will stand the wear and tear of the elements for a long time. Anyone by adopting the following formula, cannot help attaining success:

Into a tight box or barrel, put five or six gallons of hot water in which has been dissolved four or five pounds of coarse ground salt. Into this put a pail full of the best lime obtainable. The large lumps should be broken into quite small pieces. Immediately cover the barrel and cover with a heavy weight, in order to keep it in place when the lime is slaking, for the uplifting power of the boiling mass will be surprisingly great. After a few moments uncover and stir the mixture to the bottom with a long stick, then recover and keep closed for a day or two. When fully slaked the lime should be of the consistency of thick cream. When applied to hen houses or a fence, it should be thinned with water to the consistency of common paint.

If too much water is used in slaking, the lime will be drowned and as a result, the wash will be thin and watery. If not enough water is used, the lime will "burn" and granulate. If properly slaked, the mass will be smooth and free from lumps.

When applying the whitewash, dip out a sufficient quantity into a pail, then stir in a handful of cement. This will cause the wash to firmly adhere to the surface to which it is applied. It will be a dazzling whiteness and will "lay on" like paint.

An excellent plan when whitewash is to be used about the hen house, chicken coops, etc., is to put in a liberal quantity of crude carbolic acid.

This may be a lengthy description of the simple process of making whitewash, but anyone will find the recipe first-class. The old-time method of slaking lime in cold water and applying the weak solution is very unsatisfactory.

Burglar Alarm—I refer to the mention made by you of an electric burglar alarm to protect poultry houses, and would venture to inquire whether such an alarm may be installed by one not a professional electrician. Upon what principal is it based, and what are the materials needed?—H. M.

Answer—I put in the burglar alarm you speak of myself. I am not a professional electrician, but I went to the electrical supply house, bought from them the ordinary alarm fixtures which are used at the door and windows of residences; they explained to me how to set them, and I did it by their directions. I did not find it difficult. None of the doors or windows in my hennery could be opened four inches without the alarm gong at the head of my bed, ringing. I should think you would have to understand a little about it to put them in.

Moving Chicken Houses—I have an orchard and am thinking of going there and confining my hens in a house large enough for twenty-five. I propose to attach to the house a closed covered run twelve or fifteen feet by five. This house and run I could move say twenty feet every week. At that rate of movement it will take about six months going from one end to the other. The shifting of position would give new soil to scratch on each week. It would be necessary to confine the fowls as I have outlined for six or seven months to keep them away from the grape vines planted in the place. There would be some shade from young trees. When the grapes were off the vines the fowls could be turned loose, that would be during the winter months. Is the plan practical? How long do you think it might be worked until one would have to quit because of disease and loss of vitality?—H. S. T.

Answer—Your plan is quite feasible and there is no reason for it not lasting for years, but I would advise you to move the house and yard twice a week as the yard is small for the twenty-five fowls. You will have to keep the fowls healthy by making them exercise and by not feeding any mashes but plenty of green food. Moving the house and run frequently will spread the chicken droppings over the place and in a few years make your orchard very fertile. By keeping the hens vigorous, clean, out of draughts, there is no need for roup or any colds in your part of California.

Management of Poultry—Kindly let me know where I can get full information and feeding, etc., of poultry; in fact, good ideas of how to run a chicken ranch as I intended shortly to make my home in Idaho and resort to this occupation.—D. B., New York.

Answer—You had better take the Live Stock Tribune, published in Los Angeles, as it is considered the best poultry magazine west of the Rockies and deals with the conditions of the Pacific Coast and adjacent States, so that you would find this western paper of more use to you than any eastern paper. The price is only 50 cents a year.

Willing to Learn—I am thinking of starting in the poultry business and would like to ask a few questions. Are incubators a success? Why is it necessary to test the eggs? Is it best to put young chickens in a brooder or to give them to a hen? Why could one not put eggs in the incubator as they are layed, say two or three a day and take the chickens out as they hatch?—F.L.

Answer—Incubators are a success if you get a good standard make. Find out what your neighbors are using successfully. It is necessary to test the eggs to take out the infertile ones and use them for eating or cooking so as not to waste them, also the infertile egg not having life in it is cold and chills the neighbor egg which has life in it.

If you use an incubator, it is necessary to have a brooder, as you will hatch too many chickens to go under a hen.

It is not best to put eggs into the incubator as they are layed, because for the last two days of incubation the incubator should remain closed, also for the first two days—and between those periods the eggs have to be moved, turned, and taken out of the incubator and cooled, consequently it is best to save the eggs until you have enough either to put under the hen or fill the incubator.

YARD ROOM

How Many Chickens to Keep on a City Lot—Will you kindly tell me how many chickens can be kept on a city lot seventy-five by a hundred and eighty feet? Do you think chickens will lay well during the rainy season in Seattle, Wash., if they are properly fed and housed? How big a house do we need for fifty chickens?

Last September we bought thirty Plymouth Rock hens and thirty pullets. We got from 10 to 16 eggs from the hens per day, until about the middle of December when they began to fall off. We are still getting that amount, but half of them are from the pullets. Do you think they are doing as well as we could expect?—Mrs. L. E. S.

Answer—In your climate it would very much depend upon the shelter from the rain that you can give the chickens. Fifty chickens should be divided into two pens with two houses. Each house not less than ten by twelve feet in size. I would advise a good scratching pen to be made either adjoining the house and covered with a roof, or else make the scratching pen to extend underneath the dropping boards. You might keep several hundred hens upon land 75x180 feet if you have ample house room for them so they would be well sheltered from the rain. Hens that are wet every day will not lay well. Your fowls are doing well considering the wet weather you are having.

How Many on Two Acres—I have two acres of land, of which I will have a hundred feet by one hundred feet for an alfalfa patch, the rest for chickens to run around and have the patch for them to feed on for an hour or so before going to roost. Kindly let me know how many chickens I can raise on the two acres at the most.—M. J. P.

Answer—I think you can keep a thousand chickens on your two acres. You must be careful not to have more than fifty to roost in one house. It is the crowded condition of houses at night that brings trouble and disease. Be sure to give them shade during the day and plenty of good fresh water, besides, of course, the balanced ration. Allow them two hours a day on the alfalfa patch.

Five Acres—Will you kindly tell me how many White Leghorns I can successfully raise on five acres of land? I want to grow alfalfa and some vegetables for feed.

Will you also tell me if I can hatch turkeys in an incubator?—J. W. L.

Answer—You can raise a large number of Leghorns on five acres of land. I know one party that has 3,000 Leghorns on three acres, but it entirely depends upon knowing how to do and doing it right. Better begin with a small number and when you succeed with those, increase your flock.

Turkeys can be hatched in an incubator and raised in a brooder, but must be kept entirely separate from chickens, or they will die.

Yard Room—I want to raise about 60 pullets for next winter. I have about a hundred chicks hatched out. All the yard room I can spare is on a town lot about 50x75 feet. Do you think this would be enough room for them?—Mrs. J. F. Y.

Answer—It all depends upon the care you give them; if you can supply them with shade, plenty of green food, clean water and a good scratching place and the proper food, it will be plenty large enough. Be sure to keep them clean and free from mites and lice.

MATING AND BREEDING

Age for Mating—I wish to ask if a cockerel should be mated after he attains a year in age or can he just as well stay till a year and a half or two years old before being mated?

Also I wish to know if it is quite as advantageous to mate a rooster with a pullet of his own clutch, supposing the pullet and rooster are both a year and a half old. I would like to do that if you think it advisable.—M. S. H.

Answer—The earliest age at which a cockerel may be mated should be about ten months, not earlier if you want large, vigorous chickens. I consider the best age for getting sturdy chicks is for both parents to be about two year of age. You can keep a male bird as long as you wish without mating him but he should be entirely out of sight and out of hearing of the hens, otherwise he will fret to get to them. I have known several to drop down dead from getting too much excited at seeing other young males in the pens with the hens.

From a year and a half to three years of age is undoubtedly the best age at which to mate the fowls but you can have very good results with older fowls. In your place I would certainly mate the year and a half male with the year and a half hen and expect good results, for they should both be in their prime.

Mating Brother and Sister—Is there any objection to mating a rooster with hens of his own clutch if they are all old enough, say a year and a half or two years old?—Mrs. G. S. H.

Answer—It is considered best not to mate brother and sister together, yet this is always done in making any new breed, and as yours comes from a three hundred egg a year hen, I would advise you to do so.

Breeding—I have a nice R. I. R. cockerel. He is good shape and color but he is not up to standard weight. If I breed from him will he produce chicks larger than himself if they are well taken care of? Is there any chance of getting perfect specimen from fowls under weight? I bought some very fine looking hens, but their breasts are uneven. I also got eggs from the same stock and the pullets have crooked breasts. Kindly tell me if that trouble will be handed down if I breed from them—Mrs. C. R.

Answer—As a rule, the chicks take their size from the mother. If your R. I. R. hens have a good size the chickens will be larger than the cockerel, if you feed them for large frame. If the hens are under weight and size, you may have difficulty in increasing the size of the offspring. Some people think that crooked breastbones come from chickens roosting on a narrow perch when they are young; however I think it is generally conceded that crooked breast-bones are often hereditary. You will know if your chickens have roosted at too early an age. If not, it is hereditary and you had better change the strain.

Crossing Leghorns—Please let me know if it is profitable or a good plan to cross Brown Leghorn hens with a White Leghorn rooster. What kind of looking chicks are produced by such crossing?—F. V.

Answer—I do not think it would be a good plan to cross the Brown and White Leghorns. The chicks would be unevenly marked, some white, some brown, and a few mixed, and nothing would be gained by crossing.

Mating Parent and Offspring—I have a choice Barred Rock hen that I have mated with a vigorous young cockerel. I expect next year to mate her with one of her sons and the pullets with their father. Beyond this I am somewhat muddled as to the proper matings in order to establish a flock of "line-bred" fowls. Kindly explain the proper steps to take.—G. B.

Answer—It would entirely depend upon the results of your first and second mattings, the results of which nobody could possibly foretell, especially not knowing the parent birds. You would probably have to establish a double mating, keeping one pen for cockerel breeding and the other for pullet. All you can do is to wait and see the results and then get some one who is a good judge to look at your fowls and tell you which colors (whether dark or light) you should mate together, also which shapes, whether blocky or rangy.

Increasing Size of Eggs—I have a number of chicks, White Leghorns and Black Minorcas, which are penned up in a certain number in each corral, which is quite large.

I feed warm mash in the morning, at noon mixed grain, wheat, Kaffir corn, cracked corn, hulled oats and rolled barley. Afternoon I give all the cut up green clover and lettuce they can eat; before going to roost I give them all the mixed grain, the same as at noon, that they can eat. The chickens were hatched last March and in the early summer.

I would like to know if I am giving them too much clover and lettuce or too much grain, or too much of the combined food. I have very little sickness with them and get quite a quantity of eggs, only there are quite a number of small eggs layed. Will you kindly let me know if there is any known way of increasing the size of the eggs? The ones that are laying the small eggs were hatched last March.—Mrs. T. H. H.

Answer—Leghorns and Minorcas can stand more grain than Asiatics and American breeds. You cannot over-feed with green food at this season of the year. You do not mention the animal food. Each hen should have about a half-ounce of animal food per day; otherwise your feeding appears good. The only way of increasing the size of the eggs is by selecting as breeders hens that lay large eggs and only setting eggs from those fowls. Leghorn pullets lay a small egg unless they are of the "bred-to-lay" strains. The second year they lay larger eggs. A liberal feeding of animal food will increase the size of the eggs. You do not say how you make your mash, whether there is any animal food in it or not.

Buff Leghorns or Buff Orpingtons—My husband and I have read your articles with a great deal of interest for a long time. We wish to start our chicken raising on a scientific plan and are preparing to get either Buff Leghorns or Buff Orpingtons. Are the Orpingtons a good eating fowl as well as good layers? What do you think of a pure cross between the two breeds, using the Leghorn pullets and Orpington cockerels? Of course I mean to keep the cross always pure.

Where can we go to get thoroughbred fowls of either variety? Also please advise us of several good poultry farms nearby Los Angeles which we can visit. We wish to profit by other peoples' experience and save ourselves as much discouragement as we can. Are these ranches open to visitors on certain days?

Answer.—Buff Orpingtons are excellent layers and a delicious table fowl. They should commence to lay at about five months of age and are good winter layers. They lay a brown egg and are surely a beautiful bird. I do not think a cross between the Buff Orpingtons and Buff Leghorns at all advisable. There is nothing to be gained by it; nothing whatever. In making a cross, the usual way is to take a light weight male for heavy weight females. This is in order to have large chickens, as the mother very much controls the shape and size of the offspring. How would you propose to keep the cross pure? You can get thoroughbred fowls of any variety from advertisers in the Live Stock Tribune. Most of the ranches which advertise fowls for sale are glad to show them to visitors, and a good plan would be to attend the poultry shows; there you will see the best fowls of many different breeds; can there choose those you like best, become acquainted with their owners and make arrangements to visit their ranches.

QUESTIONS OF BEGINNERS

Just Starting In—1 am just starting in the chicken business and as I know very little about it, thought I would ask you a few questions.

1. Is it good to let several breeds run together if you do not set the eggs, or should each breed be confined in separate pens?

2. Is it necessary to feed meat and green food in the summer, or will they get bugs and grass enough?

3. How can I ventilate a hen house and not have a draught? Is it good to have the windows open day and night in warm weather? I have built two chicken houses with a shed bettween them and lined the houses with tar paper, but each night the chickens all go into one house, and I have to carry them into the other one. How can I make them go into both houses?

4. I have an incubator that holds 240 eggs, and would like to know how to turn the eggs quickly. It takes me about ten minutes, and I think that is longer than the eggs should be out, besides, the way I turn them (catch each egg between finger and thumb and turn it over) some of the eggs get jarred sometimes.

5. How many chickens should be in one inclosure if you want eggs? Should the young roosters be kept separate from the pullets? If so, at what age should they be separated?

6. Is it necessary to feed both ground bone and oyster shells at the same time?

7. Do oats make good feed? If so, should they be hulled?—F. A. F.

Answer—1. It is not good to let different classes of fowls run together. The Leghorns (Mediterranean Class) need a more fattening food than the Plymouth Rocks. What would make a Leghorn lay well would prevent a Plymouth Rock laying, as it would make her too fat.

2. It is necessary to feed meat and green food if hens are confined in runs or yards, and if the grass dries up, turns into hay, or becomes tough in the summer time. You must be the judge yourself about this matter.

3. You can best ventilate a house by having one side entirely open or closed only in winter or rainy weather by a burlap curtain. If you have windows, take the sashes out and replace them with burlap or leave the windows open day and night in warm weather.

When the chickens are going to roost of an evening, stand there with a broom and gently "shoo" half of them into one house and the other half into the other. This will teach them the way much better than to carry them.

4. If you will tell me the name of your incubator (each make has a differene shaped egg tray) I could tell you how to turn the eggs. Ten minuets is not too long to keep the eggs out.

5. It depends upon the size of the enclosure. Twenty-five is about the best number to keep in a colony house. Separate the young roosters from the pullets as soon as you can detect the sex.

6. Feed both ground bone and oyster shall. It is necessary.

7. Oats make excellent food for hens, increasing the fertility of the eggs and making the chicks larger and stronger. Hulled oats are the best.

Wants to Start Right—We have located in Hood River, Oregon. The rainy season commences about the first of December and lasts until about the first of March. It sometimes reaches zero, but only for a day or two, I am told. I wish to raise some chickens for the money there is in them. I just want to start with a few and see what success I have before I go in on a large scale. I think I would like the Buckeye Red or the Plymouth Rock. Which would be better adapted to this climate? I want the best winter layers. Could you tell me how to make and fasten drooping boards under the perches, as I see you advocate them in lice killing?—Mrs. C. W. M.

Answer—The Buckeye Reds and the Plymouth Rocks are both very good breeds. The Plymouth Rocks weigh about a pound more than the Buckeyes. They are both equally adapted to your climate and are considered good layers. Drooping boards are only a small platform nailed up underneath the perches about six inches below them for the droopings to fall on. They can be either slanting or level.

Best Fowls for a Greenhorn—What is the best breed for a greenhorn to commence the poultry business with? —W. H. Y.

Answer—Your question is short and direct, but a difficult one to answer briefly. If you had asked me which I considered the best all-purpose fowl, I should have answered "one of the American breeds." If you ask me which do the best with the handling of a greenhorn I must say it all depends upon the greenhorn. If you want good layers for the San Francisco market, I should advise you to get a large strain of one of the Mediterranean varieties, for they lay white eggs. As a rule the man who is unacquainted with poultry will do fully as well with that class. It is next to an impossibility to get them so fat that they will not lay. In fact they will lay on almost any kind of feeding, provided they are comfortably and cleanly housed and have some kind of litter to scratch in.

What to Do and How—As I am about to go out on a ranch where I will find 200 White Leghorns awaiting me I would like to have a few suggestions from you.

1. Is oyster shell just as good as grit?

2. What proportion and what kind of feed do you suggest for hens this time of year?

3. There are no pullets. Would you advise the forced moult for about half of the hens to prepare them for early winter layers?

4. Will hens (laying) need fresh meat scraps or bone if they have two acres of green alfalfa to run on or will the insects provide sufficient meat food?

5. Is dried meat and bone as good as fresh ground bone and meat?

6. As I have no fryers or table fowl of any size would you advise setting some hens now so as to have fowls to eat for the winter?

7. How many roosters will I need for 200 hens? I will have ten with them. Is that all that is necessary until I want fertile eggs?—Mrs. E. R. L.

Answer—Oyster shell is not the same as grit and will not take its place. Hens require both. The oyster shell is to supply the lime for the egg shell and the grit or sharp gravel to supply the place of teeth to grind the food.

2. Keep to the feed they are accustomed to have until the moult.

3. I would strongly advise the forced moult for half the hens to prepare them for winter layers.

4. Hens will not find enough animal food in our climate on your range. You will have to give them either meat-meal or dried beef scraps to supplement the few insects they will find. Or else give them the dry granulated milk.

5. This is still an unsettled question. The fresh meat scraps are better than dried if you can get them fresh and without any preservative, but if they are at all stale, or if any preservative has been used they are almost poisonous and should not be used.

6. Yes! I certainly would set a few hens and would raise some chickens with hens for winter eating.

7. You do not need any male birds at all with hens unless you want fertile eggs. Infertile eggs are considered better and keep much better than fertilized eggs. Therefore, you need not keep any male birds at all till you want to hatch the eggs. Then pen up some of your best layers with a good vigorous cockerel from an extra fine egg-laying strain and you will have fine layers for another year.

A Few Points—Will you be kind enough to answer a few questions for me concerning my chickens?

First—What causes a chicken's comb to turn black?

Second—One of my Wyandottes drooped a few days, died and upon dissecting her found she was very fat; had a fully formed egg in her, which she evidently was unable to pass. What is the trouble?

Third—Do chickens ever get too fat to lay?

Fourth—Leghorn pullets hatched last April have not begun to lay yet. Can you tell me why? I feed whole barley, wheat, bran mash, and chickens have two acres of alfalfa and clover to run on; also table scraps, charcoal, ground bone and Kaffir corn. The most of my chickens are Brown Leghorns.

Fifth—How often should chickens be fed green bone?—A New Hand.

Answer—First—A chicken's comb turning black indicates liver trouble or indigestion, usually caused by lack of green food, lack of exercise and too much starch food, or it may be poison.

Second—Your Wyandotte was egg-bound. By injecting a little olive oil and holding the lower part of her body in warm water for 20 minutes you might have saved her.

Third—Chickens frequently get too fat to lay.

Fourth—Your Leghorn pullets should all be laying by the last of November. An insufficiency of egg-making material in their food, a lack of shell or animal food and green food will keep them from laying.

Fifth—Chickens should be fed green bone every other day. If you cannot get that fresh, the dried blood and bone make a very good substitute.

Best Breeds, Etc.—I. have come from the East and am starting in chickens and would like to know which are the three best broilers, heavy chickens, and good layers.

When is the best time to set chickens? How is the best way to protect your chicks from cats? Is the scratch food you buy already mixed as good as you can make? Is there any way to cure a hen that has an egg broken inside her?—H. P. A.

Answer—In this beautiful climate all breeds of chickens do well so you had better choose those that you like best. It is like asking me which flowers grow best and which shall I plant? All do well here if you take the proper care of them. You can set a hen every month in the year here. March is one of the best months. Make a cat-proof coop to protect your chicks from cats, or keep a good fox terrier. The scratch food that you speak of is excellent. I use it because it saves me the trouble of mixing. An expert might save a hen with an egg broken inside her, but as a novice you had better cut her head off.

Beginner's Questions—I have just bought a few chickens from a woman who is going away. She told me not to feed them any bran. She had done so and they did not do well. I asked the flour and feed man about it and he said "bran was the best and rolled barley was the poorest feed." I would like to know what you advise. These two statements are flat contradictions. What is the best thing to kill mites or lice? One man told me to use lard for killing lice. He said it made the hens look pretty tough, yet it did not hurt them. I have thirty hens, all young, and some should be laying now. I feed my chickens greens in the morning. I have a grape vine and they are very fond of the leaves; I have not given them any mash; a little grain at night. You may have been asked these questions many hundreds of times, yet everyone that starts has to begin at the A B C.—H. C. L.

Answer—Bran is a very healthy food for chickens, and rolled barley is a richer or more fattening food than bran; both are good for fowls. The nutritive value of bran is 1.4 and of barley 1.6. The best thing to kill mites is to spray the house with kerosene emulsion. Burning sulphur candles is excellent if you can stop up all the cracks and air-holes of every description, then either spray the walls with water or wet the floor thoroughly and light the candles and escape out of the house shutting the door tightly behind you. Keep the house shut up for 12 hours. If the fumes of the sulphur escape from the house through the cracks, it is of but little value, therefore I prefer the spraying. For killing body lice, dusting the hens with a good insecticide is the best way. Greasing them with kerosene and lard was grandmother's method, and while it frequently kills the lice, it will make young chickens quite sick. A hen to do well, needs about six ounces of food per day. Of this one-third should be green food, one-half grain and one-sixth meat or animal food. For your thirty hens they should have twice a day, about two quarts of grain, as much green food as they can eat and a pound of meat or animal food. This is just an outline of what they should have to make them lay and keep them in good condition.

MISCELLANEOUS QUESTIONS AND ANSWERS

From Beginnng to End—I have been wanting to write you for several months, but hated to do so as I feared my letter would be too long. I would like to ask about my little chicks of last summer, so I will know what to do for them another year. I had very good hatches, but only raised about one-fourth of what I hatched. They all seemed hearty and strong when they hatched except a few which were crippled. One leg drew up like it would not bend at the knee and a few seemed to have something the matter with the cords of their neck; their heads drew back under them and they peeped pitifully. The rest grew fine until about two weeks old and some four or five weeks, when I thought the danger was over. They seemed to get sour crops and gas. I fed them a special chick food and was very careful to give them clean fresh water. They had no soft feed at all; plenty of grit. I cut up tender grass and clover and fed them when the weather did not permit them to be running out. I tried giving them a little soda and the little things would vomit, and I think every one that got that way died. What is best to do? One or two old hens seemed to be affected the same way, and I gave them a little soda water, then emptied their crops and gave a little more soda and they got all right, but the little ones were too weak. In April and May something seemed wrong with their digestive system, and they became very constipated. They were fed the same food, too. What do you think caused it and could you give me a remedy? They nearly all sickened and died. I am going to try again, for I love my chicks and hate to give them up, but am not much of a financial success with them, especially since feed is high. From my 200 hens I get only about two dozen eggs per day and feed about three gallons of wheat in the morning, then I mix a soft feed of 1½ parts bran, 1 middlings, 1 alfalfa meal, 3-4 beefscraps dried, a little salt and cayenne pepper. I feed it either dry or just dampened a little; I alternate it with egg food and then at night I feed them wheat again, three gallons or more. They always have some of the soft feed left at night. What is wrong with my feeding? The droopings looked like boiling molasses. We have a few Plymouth Rocks and White Minorcas, besides mixed breeds. What can I do to get better results? They all look nice and healthy.

I will give you the price of feeds: Wheat, $2 per 100; barley, $1.50 per sack (80 lbs.); bran, 90 cents per sack; middlings, $1.60 per sack; beefscraps, $4 per cwt.; blood meal, $4.50 cwt.; cracked corn, $2.20 cwt. Please tell me how I could best feed at these prices. I want to learn all I can. I have read your letters for so long that you seem almost like a friend. Thanking you in advance for the help I feel sure I will get, I am.—Mrs. L. D. E.

Answer—I have condensed your letter somewhat, but am glad it was long, as it enables me to judge better what you need. About the little chicks: Those that had the crippled legs were overheated in the incubator. Those whose head drew back was caused from the eggs not being properly turned during incubation, and the little ones that died of sour crops, as you call it, died from indigestion. When overheated, or not turned sufficiently, or have not had oxygen enough, the result is a weakened liver, and the yolk of the egg, which is drawn up into the bowel cavity the last days of the chick's life in the shell, cannot be digested. This yolk lies in the bowel cavity, gets hardened almost like rubber, or like a hard boiled egg, and stays there till it putrifies and poisons the chick. Diarrhoea, or "stuck-up-behind" is often one of the symptoms. The chicks, when first hatched, appear vigorous and lively, but gradually become sleepy and droopy and appear to grow smaller; are chilly or feverish and huddle together and finally die. This results from lack of oxygen and ventilation in the incubator. I would strongly advise you to change to another chick food. I think in your part of the country, large crops of oats are raised. I would advise you to get the hulled oats, or rolled breakfast oats, and cracked corn; mix and give instead of the chick food, keeping a box of beef-scrap (fine) or meat meal mixed with bran, half and half, always before the little chicks, besides feeding them plenty of green food. The constipation in April and May came from that chick feed. A little Epsom salts in their drinking water, and considerably more green food would have cured that. For little chicks, about a teaspoonful to a quart of water would be sufficient. Another thing for little chicks—they must be kept away from the older fowls; they never do well if they run together. The soda was the right treatment for the old hens, but for little chickens the damage was done before they left the egg shell, so nothing could help them. For your 200 hens I think you are feeding too much grain. A hen requires about five or six ounces of feed per day. About three ounces of grain (or bran, middlings, etc.) 2 ounces green stuff and one ounce, or less, of animal food. Now, as to your green feeds: Can you not get potatoes, or beets, or turnips, or even pumpkins cheaply? All of these fed raw are great promoters of egg laying, and are much cheaper than feeding grain. You can get machines, vegetable cutters or grinders, or you can chop them up with a common chopping knife, adding some onions occasionally, or any other vegetables. You can boil any of these and by adding bran and cornmeal and blood-meal, make a cheap and very palatable mash. It is in these little ways that you can economize in feed, making it cost very much less than when you have to buy the solid grain. If you can get oats as cheaply as wheat, I would advise you to buy hulled oats. If you cannot get hulled oats, soak or scald the whole oats and feed those to good advantage. They will make the hens lay.

The molasses-like look of the droppings comes from indigestion, brought on sometimes by too much egg food or poor beef scraps. A little charcoal and a little bi-carbonate of soda in the drinking water will usually cure this. There are a few things to remember in feeding fowls: They like a variety of grains and food. A variety does not cost any more but only gives one a little more trouble in mixing it. I get a sack of wheat, of rolled barley, cracked corn and hulled oats, mix them in one bin and feed one handful of this for each hen in the litter in the morning; or I get what is called here, scratch-food. This is a variety of grains, wheat, cracked corn, Egyptian corn, millet, sunflower seeds, etc., which I get at the poultry supply houses already mixed. I feed a handful of this every morning to each hen. I also keep before them what is called a dry mash—2 parts bran, 1 middling, 1 corn meal, 1 alfalfa meal, 1 meatmeal. This I keep before them all the time. I also give green food (lawn clippings or vegetables) and table scraps, and when the weather is cold I add a little red pepper or chili pepper seeds to this and I have eggs all the time.

Miscellaneous Questions—Give the name, price and where to obtain a good spray pump. One that a woman can use and that will spray whitewash. Will whitewash stick on the outside sacking of hen houses without any previous preparation being applied? 2. What is granulated bone fed for? How soon should it be fed to chicks? My hens eat very little and pullets will not touch it. Would it be better to mix it in the dry hopper feed? If so, how much? 3. How soon should oyster-shell be fed to pullet? Which is best for growing chicks, beef-scraps, beef-blood and bone or dried blood? If mixed in dry hopper feed, what proportion of each should be used? 4. What number of eggs a month would be considered good laying for a pen of 12 White Rocks, one year old, when laying well? Ought they to lay more next year? 5. What amount of grain should be fed night and morning to a hundred growing chickens that have dry hopper feed. 6. Would it be any advantage to always have green alfalfa in hoppers? 7. Have rhubarb leaves any value as green food?—Mrs. M. H. S.

Answer—You can get a good spray pump for spraynig whitewash for $1.25 at the poultry supply houses. You will have to thoroughly wet the burlap on the outside of the hen houses or the whitewash will run off without penetrating the sacking. 2. Granulated bone is fed principally for the phosphorus it contains. It is used for making, or strengthening the bones of the chickens, and can be fed to them from a few days of age. There is usually about from five to ten per cent. of it in chick feed, and it can be fed dry in hoppers in this proportion, or be placed before the chickens as the grit and shell is. 3. Oyster-shell should be fed to pullets before they commence to lay. It also should be kept in a box before them to help themselves as they need it for making the egg shells. I prefer dried blood to the others you mention, but they are all good, and I would advise you to use whichever you can get the easiest. 4. About 180 eggs per month would be considered good laying for twelve White Plymouth Rocks, but hens bred-to-lay will lay more than that, averaging about 270 per month for one dozen. White Plymouth Rocks with me layed their best at two years of age but I have had them lay well up to seven years of age. 5. What they will eat up. The amount depends upon their size, age and appetite. Growing chickens should be fed "full and plenty." 6. Green alfalfa should not be fed in hoppers, for if pressed down in a hopper it will soon heat, ferment and make the chickens sick. 7. Rhubarb leaves would not be desirable as food for chickens. They contain too much acid.

Advice Wanted — Would greatly appreciate a little advice in regards to my chickens.

1. How long should chicks be allowed to remain in the nursery after hatching?

2. Will you please tell me the proper temperature of a brooder containing two-day-old chicks?

3. At what age should I begin to feed wheat or larger grains to chicks?

4. When should they be made to roost?

5. What is the yearly average number of eggs per hen.

6. What should be fed to hens in the evening? —S. J.

Answer—1. Chickens should be allowed to remain in the nursery until they are dried off, but they may be left in for twenty-four or thirty-six hours if desirable.

2. I suppose you mean the heat under the hover. This should be for the second day about 90; it very much depends upon the weather and also upon the vitality of the chicks. If they are warm enough, which is all we want, they will spread them-

selves out and have a contented little song; if not warm, will huddle and crowd together, so you can soon learn to know their wants.

3. You can fed wheat mixed in the chick feed as soon as they will eat it, also Kaffir corn. They will usually commence to eat it at about three weeks of age, some earlier.

4. When you wean them from the brooder, or when the mother hen takes them on the roost with her.

5. The yearly average of a farmer hen's egg output is 100, that of egg farms is about 150, but "bred to lay" hens do better than that. I had a number that layed over 200 per year. One of them layed 267 eggs within a year. I expect we shall see a 300-egg hen some day.

6. When you tell me what breed you are handling I will tell you what I consider the best food for their supper. Each breed needs a little different handling to do its best.

Shipping Young Chicks—Do you think I can order eggs incubated 31 miles from here and have the young chicks sent by stage with perfect safety?

We are feeding corn of our own growing which is quite musty. I have been afraid of it, but so far cannot see that it has hurt them, although yesterday a hen sat around all day droopy like. I wondered if the musty corn affected her.

Last summer I brought into the house some small chicks that seemed about to die and seeing they had lice, I dusted them thoroughly with buhach. The lice soon dropped off of them, but the chickens died. Can too much powder be put on them?—Mrs. C. S.

Answer—Chickens could travel a thousand miles before they are twenty-four hours old, if packed in a box carefully. That is, of course, before they are fed. Last year I sent some from Los Angeles to Berkeley. They were out 36 hours but arrived in perfect condition, all vigorous and ready for their first meal in their new home nearly a thousand miles away.

Musty wheat or corn is very unwholesome for chickens. Buhach would not kill the most delicate chicken or turkey, but is death to all insect-life. The chickens were doubtless dying before you powdered them.

Castor-Bean Bushes—I have been thinking of planting castor bean bushes in the chicken yard for shade, but was advised by a neighbor not to do it, as the beans would drop off and if chickens ate them they would be poisoned. Would like your advice, please. The bushes grow quickly and make good shade so would like to try them. Do you think it would be O. K.—J. H. S.

Answer—Castor beans are poisonous to both ducks and chickens if they eat them, so I would advise you to plant something else. Get cutting of fig trees, about ten inches long, bury the whole length except one inch, water well and you will have shade in a few months and fruit in two years. I find figs excellent in the chicken yard and the chickens do not eat the leaves and bark. Would advise your planting also other fruit trees such as plum, peach, apricot. The chicken droppings fertilize these trees and the quantities of fruit you will have will soon repay the trouble. In the meantime you might plant sunflowers. They make good shade and their seed is excellent food for the chickens.

Capons—Will you kindly give us an article on capons? What is the demand for them, if any? What do you think of the difference in profits between them and broilers? If there is any truth in the statements published in regard to capons in the Eastern markets they ought to be money makers here. Am fitted for the business, but desire more information in that line before attempting much. I think the R. I. Reds would make extra good ones and I should like marketing mature birds instead of those of few months old. Capons for the Philadelphia market have to be a year old to command the best prices.—H. J. K.

Answer—Capons bring a good price now in Los Angeles, especially if you can make a contract with some of the large hotels for them. This you can only do by having a large and regular supply. The price last year was from 30c to 35c per pound, which is a paying price. Broilers pay about as well when you take into consideration that you can turn them off at eight weeks of age. This would be your better plan, as you are limited for space and you would not have the expense and trouble of carrying them for another ten months. I would advise you to sell as broilers all the young males you do not wish to keep for breeders. This will give you more room for the pullets and you need space to have your pullets develop well for the fall and winter egg market. Capons are, undoubtedly, money makers for those who have plenty of space, and where food is cheaper than it is here this year. Personally I found that capons did not pay as well as roasters. These were young roosters that were about eight months old and that I milk fed. I found I had to keep my young males until I could see how they would develop. I began by caponizing, but being economically inclined, I found the milk-fed uncaponized eight months youngsters paid me best. Since then the market for capons has improved here, and if you had more room and could buy up young cockerels, caponize them at about three months of age and turn them off in the following spring, just when turkeys go out, you might make some profit on them. It has been found that the Brahmas or crosses of the Brahma are the best for capons.

From Far Away Alaska—Commencing with the first of March for the last three years my chickens begin to lose their feathers in front of their neck. I feed them wheat, corn, shorts, cooked potatoes and cabbage. They have no lice. I also give them plenty of charcoal and grit. I have a chicken house 30x30, logs with moss between, lined inside with shakes. I also keep fire in a stove to keep out dampness.—H. C. C., Sumdum, Alaska.

Answer—Not knowing your climate, scarcely like to venture an opinion about the reason for your hens losing their feathers. Your rations seem good, all except there is no animal food in it. I think you should give them fish with their cooked potatoes. Do not feel alarmed about their losing their feathers as it may be on account of the climate.

Chicks Dying—I have 70 chicks three weeks old. They are fed in a hopper, cracked corn and chick feed, also have all the sour milk and field beets they will eat. They grew wonderfully fast until three weeks old, then their legs grew weak and they acted drunk. What do you think is the trouble?

I also want to ask you about my incubator. A great many of the fertile eggs fail to hatch. Some of the chicks break the shell and a sticky substance runs out around them, hardens around the edges and they die. Can you help me? How should chicks be fed for broilers and how for breeders?—S. F. F.

Answer—Your chickens are getting a too carbonaceous diet; too much corn and not enough oats, not enough green food and animal food, consequently their bodies are growing too fast for their legs to carry them. You are feeding them for broilers. If you want them for breeders, feed the chick food, adding more oats and a plentiful supply of green food and onions. Give them charcoal, also fine grit or coarse sand and they will be all right.

Your eggs apparently have become too dry in the incubator. Either the ventilation is at fault or you are not running the incubator according to directions.

___

Henpecked Husbands — I cannot keep my hens from picking the combs of the roosters. Could you tell me the reason for it? Also a remedy for it? I have tried everything I know for it. I feed meat twice a week.— R. M.

Answer—This habit or vice usually comes from a lack of green food or meat in the ration. Very often the habit is acquired by imitation and thus it may be introduced into a flock by a new bird which had contracted it elsewhere, or it is spread through the flock from a bird which is led to it by indigestion or other disease of the stomach. It is sometimes started by lice. The hen sees one crawling on her mate's comb and tries to peck at it, wounds the comb, tastes the warm sweet blood and keeps up the habit. The others imitate her until the poor henpecked husband is in a sorry plight. The preventive is plenty of green food, plenty of exercise and animal food. The cure, the hatchet for the worst hens, or if they are too valuable, let them run without the male bird, only admitting him to the pen for an hour a day in the after-noon. Give the hens a good run in a grass-covered yard. Feed plenty of green vegetables; onions chopped are particularly efficacious. If the yard is small, prepare a scratching shed, covering the floor deeply with straw and scatter grain in the straw for the morning meal, so the fowls will be compelled to scratch and work to find it. Add bi-carbonate of soda to the drinking water in the proportion of about 20 grains to the quart; put a small quantity in the food, or nail up a piece of salt pork for the hens to peck.

___

Severe Climate—Do White Leghorns stand a severe climate well?— Alberta, Canada.

Answer—White Leghorns are a healthy and vigorous bird, but the large combs of the Single Comb breed are easily frozen on account of their size, especially in a country where the temperature falls to zero or below. If a hen's comb freezes severely, she will not lay for three months. On this account I think it would be advisable for you to have the Rose Comb Leghorns, or fowls with small combs. The comb seems to be the part of the fowl to freeze the quickest, and you should use care that your fowls do not sleep in a draught during the winter.

___

Nothing in It—We expect to get a pen of White Plymouth Rocks later on, but for the present want to buy common hens, perhaps about 400 hundred. These will be Leghorns, Plymouth Rocks and mixed. We have potatoes, green vegetables, etc., corn and plenty of wild grass but would have to buy anything else needed. How old hens would it pay to buy, as I doubt whether we can get pullets and we would have to prepare to fight lice among such a collection. —D. B. B.

Answer—It would take an expert and an old one, too, to make any money out of your proposition. For a novice there is no money in it and probably a heavy loss. It is the pullets that make the winter and fall layers and if you get a number of old hens or even if they are only two year-olds they will probably not lay till next spring, and you will have all the expense of keeping them till then and the trouble of taking them through the moult, and most likely roup and colds will break out amongst them and get your whole place affected with sickness.

You would make more money by paying twice as much for pullets.

Which Breed and Which Specialty? —Am thinking seriously of trying the broiler and fryer with a capon line production, rather than an egg production, and want to know your opinion of success in making a specialty of that line.

What breed of fowl would you recommend? They, of course, should have strong vitality and mature early. Would the Rhode Island Reds fill the condition?—"E."

Answer—The broiler and fryer business depends upon what contracts you can make with the hotels. A breeder that I know, who has large contracts, gets $6.00 per dozen for chickens weighing a pound each. It costs him about 18 cents a chicken to feed and raise, not counting his own labor. He will therefore make 32 cents a piece on his fowls. Of course, this means work and plenty of it. It takes from six to eight weeks to raise a broiler; from eight to twelve weeks a fryer.

Commission men and markets do not give as high a price as hotels or restaurants. Of course, I cannot judge what would be your best market in your little city. I suppose you would have to go to San Francisco and there would be all the expense of expressage or freight to deduct from your profits. If you realize that I do not know how or where you could sell, or at what price, and that I do not know the price of food, etc., in your locality, you will see how difficult it is to give you any advice on the subject.

The Rhode Island Reds are a very vigorous fowl, and I think would suit you admirably for broilers and fryers.

Spankled Hamburgs and Broilers— (1) What do you know about the Silver Spangled Hamburg fowls? Are they worth keeping? (2) How do they compare with the Leghorns? I only know that they are pretty and I would like to have some but do not know anything else about them. I have the White and Brown Leghorns but I like different kinds if it is advisable to have them. I want to know into a larger poultry business. Would

if it be best to have only one kind, or would it be wise to have different kinds? (3) Which of the large breeds would be best for early broilers? (4) By crossing a large breed with a smaller would the chicks be ready for market earlier than if pure bred?— Mrs. B. H. K.

Answer—I have known the Silver Spangled Hamburgs all my life well, till I came to California. They are a most beautiful bird; splendid layers of a white egg which, considering the small size of the bird, are large. It is by crossing the Hamburgs with many of the larger breeds that the egg laying quality has been given to some of the recently made breeds such as the Orpington, Silver Wyandotte, etc. (2) In comparison to the Leghorn they are an older breed, being bred in England over two hundred years. They are about the size of the Leghorns but have a plumper body. They mature very early and are considered a very hardy fowl. I do not know why there are not more of them in California. It is best to have only one class of fowl but if you want to keep more you will have to fence them away from each other or they will mix. (3) The Wyandottes and Rocks are considered the best for early broilers, but the Leghorns if properly fed for broilers make very good, very early, small and plump ones. (4). No. It depends upon the feeding. If you cross breed you can never tell after which parent the young will take and you will probably have an irregular lot of young, some large, some small, some with nice plump little bodies and others with boney frames. For the market you want them as even as possible so that they all look alike, and to have them all alike you must keep to the pure bred fowls. You will find that in time the pure bred birds pay better than any others.

"Up Against It"—I am one of the very large number of chicken raisers who are constantly "getting up against it" in one way or another and rush to you for advice.. I have about three dozen chickens on the back of my lot, the range they have being 50x80 feet. They are "just chicken," having been hatched last February from market eggs purchased from the grocery. Have been feeding wheat and cracked corn mixed, also cabbage at noon, and am getting on an average about one dozen eggs per day. Have five or six old hens, age uncertain. Among these are four Blue Andalusians, three of which are laying. The fourth layed until about three weeks ago when she quit, but was apparently healthy until two or three days ago when she became droopy. I thought possibly she might have liver trouble, so doctored her for that, but she got no better, and this morning died. I opened her and found the egg bag completed filled with what appeared to be congealed yolk of eggs. This mass of matter, a lump about as large as a man's fist, was spoilt and of about the consistency of the yolk of a hard boiled egg. Can you tell me what the trouble was and what to do to prevent its occurrence in the other hens?

A neighbor of mine is also in a little trouble. She has a nice flock of R. I. Reds and also a flock of White Wyandottes. They are all apparently in fine health but do not lay. They are nine months old. They are fed clean wheat. Have charcoal, grit and shells and up to a few days ago were fed stock beets. Would the beets cause the chickens to droop and the skin turn black or dark purple? A cockerel in the above flock having been fed beets for quite a while turned sick and his comb turned almost black; some of the hens were also affected but not so bad.

One more case: A fine large hen of the above mentioned Wyandottes was found dead this morning. She was to all appearances well and ate her supper last night as usual. Upon opening her I found a clot of blood around the heart. This is probably what killed her, but what caused it? She was very fat and had a healthy looking comb. There was not a trace of any trouble other than that mentioned; neither was there any eggs.—G. B. B.

Answer—Your Blue Andalusian had what is called an ovarian tumor. This occurs, but not very often, in old hens. Treatment is, of course, impossible in these cases, as the nature of the disease can hardly be determined until after the bird's death. If such abnormal conditions are frequently found, it is an indication that there is a predisposition in that direction in that strain of birds. Your neighbor is not feeding her fowls a properly balanced ration, and I think they are getting their food without sufficient exercise. I judge this from their not laying, from the color of their combs, and from the clot of blood around the heart. This came from her being too fat, weakening the muscles of the heart so that a blood vessel became ruptured. Such a death can neither be foreseen nor treated.

The beets are beneficial to fowls and had nothing to do with their sickness. There seems to be a lack of green food and animal food in your neighbor's rations, and lack of exercise of the fowls. Let her give them a little bi-carbonate of soda in the drinking water; a small teaspoonful to every quart of water.

Geese and Ducks—I have read with interest your answers to questions in the Tribune in regard to poultry raising, so will come to you with my plea.

I would like to know what care duck and geese eggs should have when a hen is setting on them instead of the goose or duck. Also what food should they have when first hatched?

Answer—Geese and duck eggs require more heat and a longer period of incubation than hens' eggs. Five goose eggs are sufficient to place under a hen, and be sure that she turns the eggs every day or the gosling will be a cripple. The goose eggs are heavy for a hen to turn and for this reason and also because they require more heat the hen should not have more than five to care for. From nine to eleven duck eggs are the number, for the same reasons, that should be given to a hen. Goose eggs require thirty days of incubation, duck eggs twenty-eight. Hens are apt to desert them towards the last and should be watched, as they get tired of waiting for their chicks to come out. I also have had hens that were so much afraid of the queer green looking babies they hatched out that they would kill them. They seem to know that they are not proper chicks. I feed the little geese hard boiled eggs chopped fine and cracker crumbs moistened with water and sprinkle a little sand on the food. This is the first food; the next day they get the same with some lettuce chopped fine, and after this I add breakfast oats with it and bran. As early as possible I put the geese out on to the lawn; take the hen away from them and put them into a box in the wood shed or kitchen if the nights are cool, or if I am afraid of cats or other marauders. They do not require heat after a few days; sometimes not after the first day. It depends upon the weather. Geese are the easiest of fowls to raise. They are a grazing bird and must have a pasture of something green to graze on. When young they should not have whole grain, but a mash of bran and corn meal with a little animal food in it and always grass or alfalfa to graze on.

Ducks do well treated in the same way, remembering to give them a little sand with each meal.

Distinguishing Pullets—I have 18 Barred Plymouth Rock hens, some of which are too old to be profitable layers. I would like to kill them off for table use but cannot distinguish them from the yearling pullets. Can you tell me how to do so?—Mrs. G. V.

Answer—Unless you know your fowls well it is difficult to distinguish yearlings from older hens. The older hens have usually rougher legs, but not always, and their feathers are more faded than the younger; they are often heavier and they are more "bossy" in their ways, do not lay as many eggs, are lazier and fatter and moult later, and when you come to eat them you will find them tougher.

Water Glass—A few weeks ago you had an article on water glass, and I thought of trying it, but was told that it was too late, as the eggs were no good after April. Will you kindly answer in next Tribune and greatly oblige.—M. B. H.

Answer—You can put eggs down to preserve them in water glass at any time of the year, only provided that the eggs are perfectly fresh when you put them into the solution and that you keep them well covered with it. They should be kept in a cool place, such as a cellar.

Fall on Their Sides and Die—When our little chicks are about three weeks old many commence to stand upright and droop their wings and tails and fall over on their sides and die in a few hours. We give them chick feed, some lettuce or chard, and they seem to want to eat whole wheat thrown down for the hens. Grit and systershells are kept before them all the time.

Would it do to let three-week-old chicks eat much whole wheat?—C. F. L.

Answer—The feeding is all right and whole wheat will not hurt the chicks, as early as they want to eat it. It is good for them. From the symptoms you describe I think your chicks have lice and mites. These will kill them. The Mediterranean class (Leghorns) sometimes make such a vigorous growth of feathers that it seems to weigh the little fellows down and I have found that cutting their wings as close as possible wthout drawing blood relieved them from the weight.

Dipping Hens—Would you be so kind as to write and let me know about dipping hens, etc.? I have a flock of somewhere between five and six hundred. I notice some of them have lice and bunches of nits on their feathers. Whenever I have caught a hen I have greased her well, but this would take too long to go through the bunch. Is there any dip that would be strong enough and do no harm to the birds that would kill the nits with only one dipping?—W. B.

Answer—As you have so large a flock of hens and do not seem able or inclined to pull out the feathers that have nits upon them, I think you will have to dip them twice, with an interval of five or six days. The nits are sure to hatch out in about five days after they are deposited by the lice, and by twice dipping them you should get most of them. It is an excellent plan in warm weather just at the commencement of the moult to immerse the fowls in a diluted kerosene emulsion, wetting them thoroughly to the skin, or dip them in strong tobacco water, or a solution of two per cent creolin or chloro naphtholeum. A well known poultryman gives the following advice: Take the strongest and purest tobacco, 25 cents' worth being ample to clean off three hundred fowls. Make the decoction quite strong. If the user will observe a few points, no one will ever regret using tobacco to kill lice and not a solitary one will be left.

First, if the dipping is done out of doors, the thermometer should be at least 80 in the shade; second, the water should never be more than blood warm, say 98 degrees; third, and this is the most important point, every solitary feather must be made soaking wet, else you will not make a clean job of it. In dipping all fowls having heavy plumage, like the Brahmas and Cochins, the feathers must be raised with the hand and the water allowed to thoroughly wet the bird to the skin. This takes from one to two minutes for large, well feathered fowls. If a dry feather is left there will be lice upon it. Do not dip the head under, but when the fowl is quiet dip the head until all is under up to the eyes. When they will not hold still, use a small sponge and wet the top of their heads. No one who has fowls troubled with lice need fear to try this. It is very effective.

You must thoroughly clean the houses to get rid of the lice, and paint the perches with a good lice paint or liquid lice killer.

Give the hens a nice freshly dug up dust bath and they will keep themselves clean of lice. You can add one of the good lice powders to the dust bath if you wish.

Feeding Little Chicks—I will have an incubator hatch out next Wednesday. What shall I fed them? I am going to use the mother-hen brooder, and how long must I keep them in the brooder house before letting them in the yard and in the open air. The trouble I had last year was bowel complaint after a few days. They would get diarrhoea and I soon lost nearly all of them. Tell me just how to feed them to prevent this trouble and what to do if they get it.

Also tell me if cayenne pepper is injurious in any way to laying hens, or if it is good, and for what. If I sawed a barrel in two and used the preparation for preserving eggs would it be all right or must they be covered more tightly than I could fit a cover on such a receptacle?—Mrs H. S. W.

Answer—When chicks die from diarrhoea before they are a week or ten days old, it is usually caused from the eggs being over heated or chilled, one or the other, in the incubator before they are hatched. Also being chilled the first three days of their life in the brooder. This heating or chilling prevents the yolk of the egg which is drawn up into the bowel cavity of the chick the day before it is hatched out, from digesting. This gives the diarrhoea from which they die.

Chicks should not be fed for thirty-six hours after they are hatched. That is the earliest. I then feed them coarse sand and water and chick feed. The chick feed is simply a number of grains, the principal of these should be cracked wheat and steel cut or rolled oats. In most places at the poultry supply houses you can get a good chick feed already mixed. I always use this for the first three or four weeks, gradually adding to it milled oats, wheat and kaffir corn until when they are large enough to leave the brooder, their food consists of kaffir corn, wheat and hulled oats, or a good mixed "scratch food" which is prepared by most of the poultry supply houses.

Cayenne pepper is a stimulant, good to be used when hens take cold, or in damp, chilly weather and will stimulate the egg laying in the winter. The dose is a teaspoonful twice or three times a week in cold weather for a dozen hens.

About preserving eggs, see the article in this book.

Feeding in General—I have about fifteen hens and pullets and one Barred Rock cockerel. They have free range and plenty of separated milk (sweet) to drink. We have not had over two weeks of weather that chickens could not get green food. These are the questions I would like to ask:

1st. If they have free range will they get enough green food, if it is grown near, such as clover, kale, lettuce and cabbage?

2nd. If they need other animal food than the milk, what would you advise me to purchase and how late in the spring and how early in the fall should it be fed?

3rd. I see charcoal recommended; will the charcoal in the wood ashes from a cookstove be sufficient?

4th. Of what value are sunflower seeds and millet for poultry?

5th. Will cooked or raw vegetables take the place of grain or green food? I feed my chickens the scraps from the table and some grain.

6th. Will you please tell me what amount to feed, as I am afraid of over-feeding in some things and lacking in others?—Mrs. F. M. I.

Answer—1st. Yes, but when green feed is grown for chickens it is best to cut it off and feed it to them or they will destroy the vegetables by eating into the heart of them.

2nd. If they are on free range and have plenty of separated milk I do not think they will require other meat, especially if fed table scraps.

3rd. Charcoal from the wood ashes from the cookstove will be sufficient until you have over a hundred hens.

4th. Sunflower seeds are fattening and should be used at moulting time to assist the feathers to fall out. Millet is a small seed and its chief value consists in making the hens work to get it and eat is slowly. I do not like it for small chickens as it is very hard of digestion. It is also injurious to little turkeys on that account.

5th. Cooked or raw vegetables will partly take the place of green food.

6th. A Plymouth Rock hen should have about six ounces of food per day. Of this amount about two ounces may be green food and the balance table scraps, cooked vegetable and grain. I do not know what your table scraps consist of; too much bread will make the hens fat. It is a good plan to mix a little bran with the table scraps and cooked vegetables. Chopped onions and chopped chili peppers or chili pepper seeds are very wholesome for hens in moderate quantities and will increase the egg output in winter. One onion and a tablespoon of pepper seed for your fifteen hens per day, will act as a stimulant and tonic and bring more eggs.

A Variety of Queries—I must again ask your advice and a few questions.

1. Should beef scraps be given to chicks as soon as chick food is given?
2. How early should chicks be made to scratch for their food?
3. Is the White Plymouth Rock a good winter layer?
4. Is barley any special benefit to poultry?
5. What is a good ration for hens to produce a goodly amount of eggs?
6. When ought pullets to be separated from cockerels?
7. Is it good or is it injurious to have the open front of a house facing towards the west? Which is best, beef scraps or green cut bone? If the lice powder gets into the birds' eyes will it injure them?—C. W. S.

Answer—In most of the chick food that is sold at the supply stores there is beef scrap already mixed in the chick feed. You will have to inquire of the salesman and if there is none add beef scrap and fine granulated bone at the rate of ten per cent of the weight of the chick feed. 2—The chicks should scratch from the time they are two or three days old. 3—White Plymouth Rocks are excellent winter layers. 4—Barley is a good feed for fowls. It is best to have it rolled. 5—There are a number of good rations for hens. You have to remember that they need about six ounces of food per day, and that the ration should be composed of green food, animal food, and grains or their by-produces. You can either mix it yourself or get it already mixed at the supply stores. Green cut bone, if it is fresh and sweet, is the best animal food for fowls that you can give them, but it must be fresh and if you cannot get it quite fresh, a good well dried beef scrap makes a very good substitute. 6—Pullets should be separated from the cockerels when the latter become troublesome. 7—Have the open side of the house away from the night breeze and also away from the direction of the rains. Here the rain comes from the southeast and the night breeze also from the east, therefore, it is a good plan in this locality to have the house open towards the West. Lice powder would not benefit the fowls' eyes and some kinds would be very injurious to the eyesight.

FROM THE SANDWICH ISLANDS

Poultry in Honolulu—The following letter from the Hawaiian Islands interested me very much and I think a reply to the numerous questions may benefit others in similar circumstances:

Waianae, Oahu., March, 1908.

Dear Mrs. Basley:—

I have been very much interested in your poultry pages and have been following your directions in my experiments with fowls. It is about a year since I began the so-called experiments. I started with two young pullets and a cockerel which were given to me. Now I have a flock of about forty-two, besides eight I have killed for home use. During this time I lost three fowls from disease. They contracted roup from neighbor's fowls and I killed them.

I have fed the fowls wheat, cracked corn, both mixed and separately, and soft food (lawn clippings and bran) once a week. At times I use bluestone in the water. The pullets of the first setting began to lay at the age of six months. They lay an average of twelve eggs at a laying.

Different people here have gone into the poultry business but none have succeeded as far as I know. There seems to be a ready market for eggs at 30 cents a dozen and for poultry at $8 a dozen.

I intend to embark in the poultry business on a large scale in the near future and wish to obtain your valuable advice as to course of procedure.

My intentions are to raise poultry and eggs for market. I want to obtain at least five acres of land for that purpose near the railroad line and in the neighborhood of the capital (40,000 population). Of the five acres I intend to turn at least two acres into a good lawn, of an indigenous grass of these Islands, which is much relished by fowls and in fact by all animals. There are no winters here and the grass is green and insects plentiful the whole year round. The feed for chickens out here costs at an average of 1¼ cents per pound by the ton.

I intend to build coops of the open front kind. For the purpose of breeding (to renew the flock and to obtain poultry for market), I intend to keep separate about twenty hens and about two roosters.

Taking into consideration the above conditions and intentions, I wish you would kindly answer the following questions:

1. Would you advise a general purpose breed or two breeds, kept separately; one for egg production and the other for table fowls? At present I have mongrel fowls.

2. Would eggs shipped from California arrive in good condition for hatching?

3. Could I gets eggs through you for hatching?

4. Would shortage of lime decrease egg production?

5. What would you advise in regard to grit, lime and animal food (beef scraps, etc.)? Are they needed and how should they be supplied?

6. What is the average weight of Plymouth Rocks and Leghorns, say six months old? My fowls average about four pounds.

7. How long will eggs keep before spoiling?

8. Is corn meal for fowls the same as for table use?

9. What is the hopper and how is it made?

10. What is the price of eggs for home use in California? Of eggs for hatching? Plymouth Rocks and Leghorns?

11. Would you advise the use of an incubator or mother hens for hatching?

12. How are trap nests made?

13. Can one obtain on the average, as many eggs from 600 or 700 hens kept together on a run, as from about a hundred?

14. How much bluestone is needed for a gallon of water? Which is the navy bean?

15. Do roosters influence the early laying of pullets, or hens after sitting?

After answering the above questions, will you kindly outline a course you would folow in case you were here—as to kinds of feeds, breeds, etc.

Yours respectfully, F. J. Noziga.

Answer—In replying to a letter like yours, I always try to consider the questions from your own point of view, and endeavor, as far as possible, to realize the conditions of climate, soil, etc., which confront you.

Sometime ago two gentlemen, who had, for many years, lived in the Hawaiian Islands, visited my ranch and greatly admired my beautiful White Plymouth Rocks, one remarking to his father, that they would have delighted the natives in former times, and turning to me he said, "Before the natives of Hawaii were converted to Christianity they regarded White Fowls as sacred and used to sacrifice them to their gods." He then informed me that the reason fowls were so costly in the Islands was on account of the mosquitos; these annoyed the fowls so greatly that in some instances he said they caused their death, and persons who kept thoroughbred fowls had to screen their coops with fine wire screening to keep out the mosquitos. It was easier for that reason to raise ducks than chickens, as the duck feathers were closer, harder and more oily and the ducks having no combs, the mosquitos could not bite them. There were large flocks of ducks kept by Chinese in or near Honolulu.

About this time I had a letter from a lady in Hawaii, describing a disease very prevalent among hens there, and asking my advice. I diagnosed it as chicken pox, and advised putting carbolic salve on the spots. These spots, she said, looked like warts at first, but rapidly ran together until nearly the whole head was covered; she afterwards wrote me carbolic salve cured them. I think, as I thought then, that the germ of the chicken pox entered the skin through the mosquito bite, or in some way may have been implanted by the mosquito. Chicken pox was then, and probably still is, one of the difficulties in chicken raising in Honolulu. The other obstacles would probably be lice, mites or other injurious vermin.

I think your plan of using openfront houses excellent. They are proving the best kind of coops in almost every climate.

Two acres of green grass sounds good, but choose the grass which contains the most protein, the grass that is the most nourishing for the fowls, or if you have plenty of water, plant alfalfa.

To reply categorically to your questions:

1. I think you would make more money by keeping a general purpose breed, that is, one of the American class, because when their days of usefulness are past, they make large and well flavored roasts or stews.

2. Eggs very carefully packed and perfectly fresh, might stand the voyage. You had better inquire of some of the prominent breeders in Honolulu what the hatch usually has been from imported eggs.

3. Yes! When you decide upon what breed you want.

4. Yes! And I believe that both your soil and consequently your Hawaiian grasses are deficient in lime, but you could probably supply lime by shells of the shell fish or coral pounded up.

5. Grit, lime and animal food are absolute essentials not only for egg making, but also to keep the hens healthy. You should be the best judge as to how to secure these in your locality. Grit should be hard and sharp, that is, each particle should have three sharp corners; at one time I supplied the lack of grit by getting broken crockery and glass from a china store and grinding or pounding it up. The poultry supply houses now put an excellent grit on the market called "mica grit," and in many places in California, a good natural grit is to be found in gravel quarries. One load of this will last a thousand fowls more than a year. Lime is best supplied for fowls by keeping crushed oyster shell always before them, or if you cannot get that, you can mix some lime in water, to about the consistency of pancake batter, leave it twenty-four hours and then pour about a pint out on the ground in each pen. It will form a flat cake and the hens will peck little bits off as they need it. You can also supply lime by giving them lime water to drink.

About the animal food: It is rather difficult to give you advice regarding it, not being personally acquainted with your climate. Green cut bone is the best animal food that I know of for fowls. This is composed of bones and scraps of meat fresh from the butcher, ground or chopped up. It has to be fed fresh, and without any preservative being used, and here is where the trouble comes in, especially in a warm climate. The next best thing for use is the dried blood or beef scraps and dried granulated bone, or bone meal, but even here, care must be used to have them good. If they have been kept in a damp or warm place a poisonous growth frequently takes place, which kills or sickens the fowls. A noted chicken raiser says milk and red pepper will bring eggs at any season of the year. In your place I would apply to prominent chicken raisers in Honolulu and find out what they use for animal food. Possibly they may use fish.

6. The average weight of well grown Plymouth Rocks is six pounds at six months old, for Leghorns about two pounds less. There is no Standard weight for Leghorns.

7. Eggs should be incubated as soon as possible after they have been layed. They may be kept three weeks if turned over every day, but the fresher they are the better will be the hatch and the stronger will be the chicks.

8. The corn meal for chicks is the same as for table use, although an inferior and cheaper grade is often used for cattle and fowls.

9. Webster defines a "hopper" as "A chute, box, or a receptacle, usually a funnel shape with an opening at the lower part for delivering or feeding any material." The hopper used in chicken business is a box or pail to contain feed, one from which the fowls can feed at any time without being able to get at or waste the food. It is sometimes made funnel shaped and of wood, but I like best, a simple box, five inches deep and ten broad with ends cut in shape of a gable, the top of the gable being sixteen inches from the floor of the box. I make a loose roof of two boards nailed together to fit the gable. However, a tin lard pail hung or nailed against the wall and kept two-thirds full of feed will make a fairly good hopper.

10. Eggs for market have averaged 30 cents per dozen in this part of California in the past year. They are now 55 cents. Eggs for hatching are of different prices, according to the breed and the celebrity of the breeder, ranging from $2 to $25 per setting, and from $5 to $25 in incubator lots.

11. For a beginner in your climate, I would advise mother hens in preference to incubators and brooders. I think they would understand the business better than you for the first year, meanwhile, you can be studying poultry culture and taking lessons from the hen's methods.

12. Trap nests are simply next boxes with a door to them. The door closes automatically behind the hen when she steps on the nest. She remains trapped in the next box until released by an attendant. The usual plan for a trap nest is something like the figure 4 trap for trapping wild birds and pigeons. You can find good plans advertised in the poultry papers, or by going to the Hawaii Agricultural Experiment Station. The special agent in charge can let you see plans of the trap nests used at the Maine Experiment Station.

13. No! It has been proved by repeated experiments that the smaller flocks give a better return pro-rata, consequently in colony houses on large ranches fifty hens is the average number kept in each house.

14. Navy beans are the white beans used generally in "pork and beans." A small bit of bluestone (sulphate of copper) the size of a navy bean in a quart of drinking water is the right amount and will kill the germs of roup and catarrh, so preventing a spread of the disease.

15. No! Hens and pullets will lay quiet as many, some think more, eggs without a male being in the yard with them. Of course the eggs will not be fertile. They will, however, keep fresh much longer and are superior for market purposes.

What course would I follow—, If I were in Hawaii and intended as you do to go into the poultry business? I would be inquiring at the Agricultural Experiment Station, and also at the principal markets, decide which breed of fowls in Honolulu would bring me in the most money, as a market fowl. Also whether colored or white eggs are the most desirable. I would attend the poultry show, observe the birds, decide which I prefer, talk to the exhibitors, learn the methods of the successful men, and follow in their steps, until I knew the business so thoroughly that I could run without assistance. As for the feeds to use, I would say most emphatically, consult the director of the Hawaii Agricultural Experiment Station, or send there for Bulletin number 13, of that station.

TURKEY QUESTIONS AND ANSWERS

Tomatoes for Turkeys—I am feeding my turkeys a small ration of ripe tomatoes. Is this a proper food for them?—W. F. G.

Answer—A small amount of ripe tomatoes will not do your turkeys any harm. They are very fond of them, and it will benefit them, although there is very little nourishment in the tomatoes.

Turkeys Have Chicken-Pox—What is the matter with my young turkeys, and what shall I do for them? All over their heads and bills there are lumps forming like warts. Some of them have just a few while others have their heads covered with them. The turkeys are about half grown. They are not penned up and have plenty of green alfalfa. We feed wheat and meat scraps occasionally.—Miss M. M.

Answer—Your turkeys have chicken-pox. The cure is to apply carbolic salve, or carbolated vaseline. In three days bathe the affected parts with warm soapsuds in which are a few drops of carbolic acid, and again apply the salve. Add a little sulphur to their food. This will hasten the cure. They should be cured in a little over a week. Be sure to separate all the fowls affected from the flock. This will prevent the spreading of the disease.

Turkeys Dying—What is the cause of turkeys two weeks old having feet crippled and crooked, and then dying? Will egg-shells given to chickens cause them to eat eggs?—A. S.

Answer—Turkeys getting out into the wet or damp grass in the morning or being in damp, filthy coops, will cause their feet to become crippled. Egg-shells given to little chickens will not cause them to eat eggs. For old hens, egg-shells should either be baked or dried and pounded up.

Running Bowels—I have a fine turkey gobbler that is suffering with running of the bowels, doesn't eat and is quite droopy. Will you kindly suggest some remedy? It is a valuable turkey.—Mrs. A. W.

Answer—Give the turkey a liver pill such as you use in your family, or a grain of calomel. Follow this with a grain of quinine given every night for four days. Feed rice boiled in milk, adding to every pint one tablespoonful of ground cinnamon.

General Care of Turkeys—I would like to ask a few questions about turkeys. You mentioned raising them in a brooder. 1. How warm should one have the brooder when the poults are first put in? 2. At the end of the first week what should the temperature be lowered to? 3. Where should we hang the thermometer? 4. Is alfalfa meal necessary or of any benefit to little poults or to little chicks if they have all the green barley they will eat cut fine?—A Beginner.

Answer—The heat under the hover should be about 95. The reason I say "about" is that on a very warm, sunny day it might be a little lower, but should the outside temperature be cold or the weather damp and gloomy, it might be up to 95 for the best results. 2. About 85, depending somewhat on the outside air and weather. Gradually lower the temperature till you get it to 70 or to 80, according to the weather. 3. There is a hole in the cover of your brooder for the thermometer to hang; keep your thermometer hanging there in its proper place. 4. No! Little turkeys require the succulent green, not the dried hay, ground up. Give them lettuce chopped up at first with every meal; then either lettuce, dandelion leaves, onion tops chopped fine, or cabbage or the tender leaves of beets. Any green vegetable that you would eat yourself will do and also the green barley as long as it is succulent and tender. Barley soon gets tough and hard and then is not suitable for the little turkeys.

Keep Separate from Chicks—Will you kindly give me some information concerning newly hatched turkeys? We have two hens and a tom. Would you advise keeping them away from chickens?—Mrs. C. B.

Answer—Little turkeys do much better when kept away from chickens. They require, or do better, on different food, and when very young require to be kept quiet, whilst the chicks like to scratch and rustle. Turkeys move more slowly and need rest and quiet. Then, again, corn, Kaffir corn and corn meal suit chickens, but ferment inside the little turkeys and give them diarrhoea, which is often fatal. Let the turkey mothers take care of the little turkeys and give them grass or alfalfa to run on and they will do well.

Lumps on Their Head—I have been trying to find out what ails my turkeys. First, my young turkeys, about three months old, have hard lumps all over their heads, and some on their legs; they do not act sick; they eat well and their heads are not swollen. Then, the old mother turkey has her head swollen in front of the eyes, soft and puffy; eyes mattered, nose running, and she acts as though she could not see to pick up the feed off the ground, although she acts hungry. They have free range on alfalfa, have free access to a trough of separated milk, and are fed wheat. They roost on top of the hen houses in the open air. I hope I may hear from you before any more of my turkeys are affected.—A. I. R.

Answer—Your turkeys have chicken-pox. Put carbolated vaseline on all of the lumps twice a week until cured. Give all of the turkeys a two-grain pill of quinine every night for a week. Treat the mother turkey in the same way. Also up her nostrils squirt from a sewing machine oil can the following: Mix two teaspoonful of castor oil, one of turpentine, one of coal oil, one of camphorated oil and four drops of carbolic acid. Shake before using; also insert this in the cleft of the mouth. If you give the turkeys chopped onions, it is an excel- lent tonic for the liver, and will prevent their taking cold and hasten the cure of the chicken-pox.

Breeding Turkeys—Your advice regarding my ducks was so good that I am going to trouble you about another matter, but first I must tell you I have followed your advice, and my ducks are doing fine, and appear to have fully recovered. I bought some turkeys last spring, four hens and a tom, and as I had never had any poultry of that kind, I did it by way of an experiment. I did very well for a beginner, and as I heard it was bad to inbreed I sold the tom and bought another about a week ago. It is a young tom hatched early last spring, and now they tell me I will not be able to raise any turkeys from him, as he is too young. I paid a big price for him and am not able to afford to buy another, so am in something of a quandry. Will you kindly tell me if I have made a mistake in buying a young tom? My turkeys are the common black kind, and there are very few that keep turkeys around here, so that I had some trouble to get the one I have.—Mrs. W. D. W.

Answer—It was a pity that you sold your tom, especially as you had such good luck with his off-spring. It would have been better to have kept him than to get a very young tom. However, if your young tom is well developed and vigorous, you will be able to raise turkeys from him. You say he was hatched early last spring. In that case he should have a fairly well developed beard, a tuft growing on his breast. When this is developed, it is a sign that he is mature enough to be mated. If he is too young, the eggs will be infertile, or the poults will be small and perhaps delicate. He may make a good breeder for next year.

Turkeys with a Bad Cold—I would like a little advice in regard to my turkeys. I have two which seem to have a bad cold. Their nose fills up and their heads swell up in front of the eyes so they can hardly see, and they are constantly making a cough- ing noise. They have a good appetite and do not seem to lose much flesh. They were very heavy when they took sick. They are the Bronze tur- keys and full grown—Mrs. S.

Answer—Your turkeys have a bad cold, which may turn into roup. Mix the following and squirt a drop up each nostril and in the cleft of the mouth. I use the small sewing ma- chine oil can. Mix two teaspoonsful of castor oil, one each of turpentine, kerosene and camphorated oil with four drops of carbolic acid. Also give the turkeys every night for a week a two-grain pill of quinine. Add chop- ped onion and a teaspoonful of sul- phur to their mash at night.

A Lack of Green Food—I have a tom turkey that is sick. He was a year old last May and about six weeks ago he would not eat. He did not look sick, and would strut and gobble a little, but did not eat. I gave him Carters' liver pills and he soon got all right. About a week ago he began to get off his feed again, and I at once began to doctor him. Have given him liver pills and germazone, but he has not eaten anything since last Wednesday. Can you tell me what ails him and what to do for him? He is a very valuable bird and I am anxious to have him get well. His usual feed is bran, barley meal, alfalfa meal and beef scrap in the morning and wheat and Kaffir corn at night, with plenty of grit and oyster shell.—Mrs. G. H. B.

Answer—I think your turkey requires more green food than you are giving him as you only mention alfalfa meal. Give him now, a quinine pill (two grains) every night for a week. Add charcoal and chopped onions to his mash in the morning, and plenty of green food once or twice a day. Give him as large a range as possible, or if you cannot give him range, let him out on your own lawn for two hours before sundown. What he needs is fresh green food and chopped onions for the liver tonic.

Sick Gobbler—I write again in regard to a fine gobbler. He was hatched last May. He has been sick about ten days. Just sits around and does not walk much. Eats very little, and his droppings are nearly all white and small in quantity. His food has been rolled barley, wheat, and we have nine acres in green barley. He has plenty of clean, pure water and is not lousey, as I dust my turkeys with insecticide every week. When he first drooped around I gave him some liver pills, but he does not get much better. I hope you may be able to tell me something that will help him as I should feel very badly to lose him.—Mrs. S. H. J.

Answer—I would advise you first to stop dusting that gobbler with insect powder, as it may be disagreeing with him. Secondly, I would give him small liver pills, and at the same time, for at least a week, a pill of one or two grains of quinine every night. Also notice his droppings, if possible, because he may have intestinal worms, although the symptoms are more like kidney trouble.

Sore Foot—I have a turkey hen with a sore foot. I think she must have sprained it flying off the roosts. I have tried several remedies, but none of them seem to do any good. It is very feverish. Please tell me how to treat it.—Miss A. E. C.

Answer—Bathe the foot with the following mixture: One cup of vinegar, one cup of turpentine, one heaping tablespoon saltpetre. If the turkey is feverish, give her one drop of aconite in a tablespoon of milk. If the swelling is caused by a stone bruise and it festers, you will have to lance it and keep it well dressed with peroxide of hydrogen, then sprinkle iodoform on it and keep it bandaged. If it is only a sprain the first will cure it.

Tapeworm in Turkeys—I have over 100 turkeys that seem to be healthy but do not grow as they should. I find now they are full of long worms, probably tape worms. What shall I do?—Mrs. L. B. D.

Answer—If your turkeys have tapeworms the best remedy I know is male-fern (felix mas). It may be used in the form of a powder; (dose thirty grains to one dram) or of liquid extract (dose fifteen to thirty drops). It should be given in the morning and evening before feeding. Oil of turpentine is an excellent remedy for the common round worm; dose one to three teaspoonsful in an equal amount of castor oil. Feeding stewed garlic or raw onions will help the cure.

Sick Turkey—I have had rather bad luck with my turkeys. I opened one after it had died and found its appendix, if such they have, full of a cheesy substance, and when I opened that it smelled very bad. You have probably had such cases to deal with, therefore I thought perhaps you could enlighten me a little and enable me to save some of them.—Mrs. W.

Answer—It is the same trouble I have written about very often. It is caused by wrong feeding: too much starchy food, grain, corn, etc., and an insufficiency of green food. Dr. Cushman recommends "something sour and something bitter." I give a liver pill, followed by quinine and sour milk, and keep them on green food only, giving them no grain at all until they recover. Of course, if the disease is allowed to go so far as to fill the bowels with cheesy matter, there is no possible cure, but at the very first symptoms, that is, when the turkeys move or walk slowly, seem dumpy or sleepy, by commencing then with something bitter and something sour, they can be saved.

Need a Tonic—I have a pair of Bronze turkeys, have good range for them and they roost in the hen house with about three dozen chickens. My turkey hen was taken sick by being droopy, acted like she had rheumatism, would not eat nor walk around. I gave her a little liver pill each night for four nights after which she seemed quite well; fed green stuff and a small ration of oats and wheat and now she seems to be nearly blind so I have to hold a dish up as high as her head for her to see to eat. She cannot pick from the ground. The droppings at first were thin and whitish. Is there anything that can be done for her eyesight?—Mrs. Z. M. B.

Answer—I am sorry about your turkey hen. You do not say if the eyes appear inflamed, or if they are running or what other symptoms there are; I think, therefore, that it is the same old trouble and I would advise you to give her a one-grain pill of quinine every night for a week. It is not good for turkeys to be shut up in a house at night and especially with hens, for they need plenty of fresh air, also if the hens have lice or mites they go to the eyes to drink and they must be affecting the turkeys. Give plenty of green food and every day an onion finely chopped. Onions are a very good liver and kidney tonic and the white droppings indicate the kidneys are affected.

Shipping Turkeys—Can turkey eggs be hatched successfully in an incubator or are they more apt to die? Will it hurt the little turkeys to be carried on the car any great distance? —Mrs. A. P.

Answer—Turkey eggs can be hatched in an incubator, if you don't mix them with other eggs, otherwise they do better under the hen.

They can be raised in brooders, and it will not hurt them to travel on the cars if they do not get chilled.

Hatching Turkey Eggs—Would like more advise in regard to turkey eggs. Should the temperature be higher than one hundred and two degrees and must it stand at that all through the hatch? How often must the water be drawn off? Is it best to leave them in three days before turning? Should the lamp chimney be washed with water? Should I keep the windows and doors open in the room for proper ventilation, and especially at night as it is cool and the temperature is hard to keep at 102 degrees? I have kept the windows and doors open today and the thermometer stays at 102, but at night it falls as low as 90 degrees. The room registers 50 degrees at night.— Mrs. A. P.

Answer—If you had told me the make of incubator you are using I could have answered your inquiries more intelligently. Turkey eggs being larger than those of the hen require a slightly lower temperature as a rule. The water in the tank, or pipes, if you have hot water machine, should not be drawn off at all. The pipes should simply be kept full. Leave the eggs in 48 hours before turning them. The lamp chimney can be washed with hot suds or any thing that will make it clean. You do not say if it is tin or glass.

The air in the room should always be fresh but the temperature of the room should not vary greatly, therefore a cellar is a much better place for the incubator than a room, and a room on the ground floor is better than one higher up, both on account of the difference of temperature and also on account of the vibration in the building.

If the temperature in your incubator varies greatly, you certainly will make a failure with turkey eggs, and had better not attempt to hatch them in an incubator but use hens for that purpose. The temperature must be kept perfectly even and not vary more than a half degree between night and day in the incubator. I think 102, if it is a hot water machine, will be about right to start it, but that depends upon where you hang your thermometer. Is is placed on or level with the eggs, or how high above them?

Cold in the Head—Will you kindly tell me what I can do to cure my turkey? He has had a cold in the head for over a month; foam in his eyes; it is clear and does not smell bad; bowels all right. I have used your mixture of oils (castor oil, coal oil, turpentine, camphorated oil and eight drops carbolic acid) in a machine oil can, a drop in nose and cleft of mouth. One quinine pill, one teaspoonful sulphur and a half teacup of onions at night for one week. This cured my turkey hen and also my chickens. The tom is no better. He eats good, struts some, but gets poorer. I have an old cellar in the ground with dirt roof for a coop and keep turkeys and chickens in it. Do you think I had better keep turkeys in a warm shed or stable? We have a very cold wind blowing all the time. I feed meat, wheat, oats, barley, cabbage and other vegetables all they want; charcoal, grit and water with germozone or bluestone in it. Please answer as soon as possible and oblige.—Mrs. E. R.

Answer—You had better continue the quinine for a week each night. Squirt up each nostril and in the clerf of the mouth one drop of the following: Peroxide of hydrogen, one part, two parts water. Do this twice a day for a week, then use the oils on alternate days with the peroxide if he is not well. Continue the onion, and if you have milk give him some. Examine for lice and keep the turkeys separate from the chickens. Turkeys do much better when allowed to roost out of doors or where they have plenty of fresh air.

How Many Toms?—I want to ask you how many turkey toms I should have for 24 hens. I have two fine toms weighing about 22 pounds each. Their beards are well developed and they appear to be very good birds. Will those two be enough for 24 hens?—Mrs. C. B. L.

Answer—It really would be better to have three toms, but under the circumstances I would rather risk having two good toms than to buy a third of unknown quality.

The rule is one yearling tom to ten hens. One tom will do for twenty hens some times but ten hens is about the best number.

Little Turkeys Dying—I had eight little turkeys hatch the last day of the year. They seemed to be doing nicely until these last two weeks. I brought them into the house as we were having lots of rain. Since then I have lost all but one. I will notice them drooping a little and in a short time I find them dead. They eat up to a short time before they die. In the morning I give them a little crushed Kaffir corn, and a little chopped onion; at noon a little curd and at night Kaffir corn—Mrs. E. G. D.

Answer—It is I think wrong feeding that causes your loss of turkeys. Corn or Kaffir corn ferments in the little gizzard and is the cause of all the trouble. Of course, when the turkeys were outside they could pick up a little green, but shut in the house they were obliged to eat the corn. Many years ago I gave up feeding corn to young turkeys. In an article in this book I give the way I feed baby turkeys and I do not hesitate to promise you that if you follow exactly my methods you will be successful with turkeys.

Feeding Corn—I have some fine Bronze Turkeys hatched May 1st, and raised by your directions. They are beginning to get quite red around the head and neck, with the skin near the top quite blue; they are almost as high as they will be. Is it safe to begin feeding them corn yet? I feed now a mash of alfalfa meal and bran, mixed with milk most of the time and rolled barley and wheat, with what green food I can.—Mrs. M. C. D.

Answer—If you want to add corn for fattening your turkeys, you can do so by adding by slow degrees, corn meal to the mash. It is the alfalfa meal and the green food that is keeping your turkeys healthy, so do not give up these. You can add cracked corn to the rolled barley and wheat, adding only one handful at first and after a few days, two handfuls, etc.

Beginner's Questions—Am thinking of trying to raise turkeys, the white kind; have plenty of room for them and very nice range, and think this dry climate would be an ideal place to raise them. Would you advise trying to raise them without any brooder providing we have no hen ready to set when the eggs are received?—R. C. M.

Answer—I think you would find the White Holland turkeys do well in the climate of Tucson, especially if you follow my directions for the feeding and care of them. I would not advise you to try hatching them in an incubator, because they do not hatch well with common hens eggs. Better borrow or buy a hen and be sure you sprinkle her with buhach powder every five days while she is sitting, so she may be entirely free from vermin when the little turkeys hatch out.

———

Liver Trouble—We are in trouble with our little turkeys, and would like to ask you to help us. They were fine, strong fellows until a few days ago, when four of them suddenly died. I just noticed two of them, a little drooping in the afternoon, and four were dead the next morning. There was the slightest touch of diarrhoea noticeable, and I immediately put a little germazone in their water, and they have had it for several days. They have no signs of it now but four more died last night, and several others are drooping. We made an examination this morning and found the liver all blotched and spotted all over in dark rings. That is all we could find wrong. The gizzard was healthy and full of grit and seemed perfect and in order.—Mrs. A. H.

Answer—The spotted liver is all that killed them. It denotes congestion of the liver. This is usually brought on by wrong feeding, or overfeeding, but it also comes from their taking cold; either from being too warm at night, under the chicken hen, getting then hot and sweaty, and then coming out in the morning into the cool, foggy air, which gives them a sudden chill. This would affect the liver, and make even the proper food disagree with them. They may take cold and get a chill affecting the liver, from running in damp alfalfa; or the chicken hen may drag them about and make the exercise too much, and this also would weaken their liver and make them susceptible to cold, which would affect their liver. I can only give you these suggestions, as I do not know all your conditions. One of the best remedies for diarrhoea in both chickens and little turkeys, is rice boiled in milk, with a tablespoonful of ground cinnamon to every pint of milk. Rice given even dry will help in a case of this kind.

ABOUT DUCKS

Wrong Feeding—I raised some ducks last year and now just about the time they should lay they take sick, sit around, don't eat, and their droppings are green. My drakes are never sick but the ducks all go the same way.

I just bought a little duck and today she is ailing, so I gave her asafoetida and some red pepper and thought that perhaps might help her. —Mrs. E. L. D.

Answer—Ducks are not subject to diseases like chickens so I feel sure that you are not feeding them rightly. You are probably giving them whole grain instead of ground grain and most likely they do not have coarse sand to eat and help them digest their food. Feed according to the directions I have given in this book.

Ducks and Rocks—I wrote you about two months ago about my intoxicated duck; you informed me that the hissing sound was a good indication it was a drake, which it proved to be, and your receipt for indigestion cured it. Now it is my fine flock of White Rocks. Last week one commenced to droop its wings and act stupid. A man was here today and says it is roup. I do not think that it is, as there is no discharge from the nose and no swell head; eats heartily, only drooping wings and takes long strides when walking. I notice another is dumpy today. Do you think it is lice or roup? I think it is lice, yet I cannot find one on them. Now, about my ducks: Does it pay me to keep them? You know how ducks eat; they were hatched in February but I am not getting an egg.—Mrs. J. H. B.

Answer—Drooping wings is almost always a sign that the lungs are affected; also I think your pullet probably has lice. These run so fast it is hard to see them. She may also have a little cold and bronchitis. Dust and fumigate for the lice, and give her a teaspoonful of honey and a quinine pill for the bronchitis every night for a week.

About your ducks: If they are Peking, they will probably begin to lay in December. When once they start in they are pretty good layers. Ducks require more animal food that chickens. You will have to decide for yourself whether it will pay you to keep them or not. If they can have plenty of green food and plenty of animal food, besides sand and plenty of crushed oyster shell, they will lay.

Duck Eggs vs. Hen Eggs—What difference, if any, should there be in running an incubator with duck eggs from hen eggs? I am very successful with hen eggs but never succeeded very well with duck eggs the same eggs hatch 90 per cent under a hen, and the first test from the incubator is about 90 per cent and then they die in the shell.—J. W. L.

Answer—Duck eggs require different treatment than the hen eggs. After the first test when you take them out to turn them, sprinkle them every day with warm water. Leave them out a few minutes to partially dry off, fan the stale air out of the incubator and then replace them. By this means I think you will have a better hatch. Duck eggs require more drying out than hen eggs and yet the shell must be dampened to make it brittle. Putting water into the incubator does not do as well as sprinkling.

Food—Good and Bad—1. Would lettuce make good greens to sow in runways for Indian Runner ducks?

2. Will some whole wheat hurt them if they are provided with grit?

3. At what age should ducks hatched in March commence laying?

4. Will beef suet and chopped fresh beef do to feed them?—Mrs. F. H.

Answer—1. Lettuce is good for all fowls and would be good for the ducks as long as it lasts, but I am afraid the little fellows would soon pull it all up.

2. Whole wheat is not as good for little ducks as bran and cornmeal. See article in this book.

3. Indian Runners hatched in March will commence laying in September.

4. Beef suet is not the food for

ducks, but if you want to fatten them, you might add a little of it to their mash.

Incubator Ducks—We want to know the proper way to operate an incubator to hatch ducks. I have had fairly good luck hatching chickens but not with my ducks. I got only 40 out of 112 fertile eggs, and this time we should like to have a few directions to go by.

Do they require as much as chickens as to moisture; do you sprinkle, also how often, and as to airing the eggs, what time of day and how long do you advise to leave the machine open; how often do you test the eggs?—Mrs. W.

Answer—Duck eggs require quite as much heat as those of the chickens; they require more airing. Should be sprinkled with warm water once the first week, twice the second and every day thereafter, but do not put any water in the pans. Sprinkling the eggs helps to make the shells more brittle so the ducks will get out easier. Test the 5th day and again about once every week to take out the dead germs, as they putrify and are injurious to the rest. When you air the eggs, which you should do twice a day, that is every twelve hours, fan the stale air out of the incubator and then close up. Commence to air the eggs when you commence to turn them, that is 48 hours after they have been in the machine. The air space in the egg should be at the large end. I think if you follow the directions from the maker of the machine, and these hints, you will have a good hatch.

Indigestion—What is wrong with my ducks? They are almost full grown, and they turn over on their backs and are unable to get up; they are very weak; their eyes scale over and some of them have died. They act very much like chickens with the roup, only they do not swell around the head.—Mrs. J. G. C.

Answer—Your ducks are suffering from indigestion and also from their heads being stopped up. The indigestion comes partly from their not having sufficient sand with their food, and their heads being stopped up, comes from the drinking vessel not being deep enough so they can rinse their nostrils out many times during the day. If you remedy these two causes of trouble in the duck yard and feed them properly, giving but little whole grain, I think they will soon recover.

A Good Ration—We have thirty ducks, Indian Runners. They all quit laying the middle of July. They are fed bran, corn meal, beef scraps and sand, in a dry mash; they have shells, charcoal, and twenty acres of alfalfa to run on, and several times a summer, an irrigating ditch to swim in. It seems to me they should not stop laying until September or October. Now, why would not Mr. Fox's recipe start them? If you remember, it consists of the following: Take 10 ℔s. bone meal, 10 ℔s. beef scraps, 5 ℔s. fenugreek, 2 ℔s. sulphur, 2 ℔s. charcoal, ½ ℔. cayenne pepper, ½ ℔. salt; mix and keep; put half pint in the mash every morning for 20 hens. It works to a charm with our Leghorns. I fed it all last winter and up to the present time, only leaving out red pepper in summer. Would the sulphur and red pepper be injurious to the ducks?

One more question: You say don't feed yellow corn to white hens. I cannot get white corn, and oats are too high. Now, what is the matter with bran, rolled barley and Egyptian Corn for hens?—N. E. M.

Answer—Sorry to hear that your Indian Runners have gone on a strike. Indian Runners usually rest in September and October, making up in the other ten months of the year with great prolificacy. You had better let them rest for their two months, and then try the effect of Mr. Fox's tonic. It is a good one I know, and the red pepper and sulphur will not hurt the ducks, although if the weather be warm, it would be better to halve the quantity of red pepper.

The Egyptian corn would do well in replacing the yellow corn. The analysis of the two are very much alike, but there is not as much of the yellow pigment in the Egyptian corn, and I would advise it instead of the common Indian corn. Bran, rolled barley and Egyptian corn would make an excellent ration.

Two Questions—I enjoyed your article on ducks, but it did not cover all I wished to know. Do ducks need to have water more than for drinking purposes? How many ducks to one drake?—D. B.

Answer—Ducks do not require water except for drinking purposes, although they enjoy a bath occasionally as much as we do.

With Pekins, five ducks to one drake is the rule. With Indian Runners one drake is sufficient for ten ducks, and they do better with this number.

Weight at Ten Weeks—Will you please inform me what weight most of the duck men can put on Indian Runner ducks at ten weeks?—I. L. R.

Answer—Indian Runners at ten weeks of age weigh as much as do the Pekins at that time, namely, about eight pounds per pair. They should be sent to market at from eight to ten weeks of age. After that, the pin feathers develop, making them very hard to pick. I think you will be greatly pleased with the ducks when you try them. Their flesh is very delicious, fine grained and the bones are small. They have very much the flavor of the canvas-back, and I have heard, are sometimes sold instead of them. They are also the greatest layers of any known fowl; the eggs are white and very delicious, with no strong taste like the eggs of other varieties of ducks.

To Secure Fertility—I am starting to raise Indian Runner ducks and want to ask you how many ducks to put with one drake of this variety, so as to secure the highest possible fertility of eggs without keeping unnecessary drakes? I have a flock of 20 ducks and within a few days will be ready to start my incubator, so if you will kindly reply as soon as possible, I will be very much obliged to you—L. F. R.

Answer—The number of Indian Runner ducks to one drake is 10. This has been found to be the best number for Indian Runners, although you can mate fifteen ducks to one drake and have good fertility. I want, however, to warn you that the eggs are not nearly so fertile in the Fall and Winter as they are in the Spring, so you must not be disappointed if at least half of the eggs are infertile at this time of the year. To increase the fertility, would advise you to increase the amount of animal food you are feeding. You can tell in five days of incubation whether the eggs are fertile and those that are not fertile should be removed from the incubator and can be used for cooking or eating. They are merely infertile eggs that have been kept in a warm place for five days, and are better than most store eggs.

Feeding for Eggs—I bought some Indian Runner ducks, thirty-six in all, and six drakes. They were laying up to the middle of December; since then that have layed none. I feed them about everything that would come from a first-class hotel, —bread, meat, oat and corn-meal mush, all kinds of vegtable and fruit. Three times a week I mix cracked corn and bran. I feed in the morning, twelve quarts, same amount at night. They have access to plenty of running water and keep perfectly clean. The pen is covered with forest leaves that makes it warm. What I want to know is, am I feeding right for laying later on? Is it customary to pick them? Does it affect their laying? I have over two hundred eggs engaged at 10 cents a piece. I want to raise all I can the coming season.—J. W. A.

Answer—I think that your hotel waste may have rather more bread in it than is good for egg production. Indian Runner ducks usually stop laying in October, commencing again in December, and getting into full lay in February. The best time for hatching Indian Runners is from the first of February to the end of July; the eggs are very fertile at such time. It may be that you are fattening the ducks too much, as over-fat ducks do not lay well. They require much more animal food than chickens. In their wild state they live on grasses, fish, frogs and insects, with but very little grain. If you think they are getting too much bread, you might save some of it for chickens, and increase the amount of meat; keep them well supplied with coarse sand, grit and crushed oyster shells.

Picking the ducks affects their laying, and it greatly prevents the drakes from being fertile. While they are moulting the eggs are never fertile.

Geese—I have a few geese and just lately they have started to lay; gather from four to six daily. Do you think by turning them daily I might save them up for incubation? About what degree should be kept up for them? I put seven eggs under a hen. Would you also tell me what should baby geese be fed?—J. W.

Answer—You can keep geese eggs, by turning them every day, for three weeks. They take thirty days to incubate. The incubator should be about 102½ for the first week and 103 afterwards. Five eggs is plenty to put under a hen. See instructions in this book for hatching duck eggs in an incubator. Treat goose eggs in the same way. Feed baby geese the same as baby ducks for the first week, gradually adding chopped lettuce until at least half their food is green food. Geese are grazing animals and require plenty of green, succulent food. They are very easy to raise and do not require brooder heat more than a few days.

Ducks, Turkeys, Geese—I have read so many articles from your pen and they have been of so much help to me that now that I am in trouble I will be very thankful if you will give me some advice.

I had some little ducks that were hatched a week ago. I fed them next day bread crumbs with a fountain of water near; afterwards on cornmeal mixed with water and baked, then moistened with milk. They had plenty of water to drink. I have to keep them in a box at night, as I took them away from the turkey they were hatched under. When four days old on taking them out in the morning one dumped around and would neither eat or drink. It breathed hard like a child with fever. It lived twenty-four hours and died. I now have them in a pen where there is a shallow ditch and running water. Another was taken the same way this morning and died at noon. I found a solid yellow substance, not overly hard, inside of it. I took it to be the yolk of the egg.

I would like to know where the trouble is and what I did that I should not have done; also what can I do to save the balance of them? I have three hens to hatch in about a week, and I do not want to make the same mistake again. How soon after hatching is it safe to put the little ducks in the pen with running water? Can I hatch duck, turkey and goose eggs in the same incubator together? About how much and how often should little ducks be fed?—M. V. A.

Answer—You have made one great omission in feeding the little ducks. Always give the little ducks sand sprinkled on the food they eat. This is absolutely necessary. Your little ducks seem to have been chilled, either from the water being too cold or from their taking cold from being removed from their mother. In this climate ducks only require heat for a short time, but they should not be allowed to get into water until they are eight weeks old, or they will get the cramps and die.

Geese eggs do not do well in an incubator, and I would not advise your putting them in with turkey and duck eggs.

Little ducks require to be fed five times a day at first. I feed them bran and rolled oats with chopped lettuce, and always a teaspoonful of sand. Give them water to drink all the time in a vessel deep enough to get their whole bill under, as they have to rinse their nostrils each time they eat, but not deep enough for them to get into.

Breeding Rocks and Geese—I have a flock of Plymouth Rock chickens: fourteen hens and two roosters; now, what I want to know is, would you recommend me to get a different rooster, as both of my roosters are related to the hens? They are all from the same pen, and are pure blooded stock. I do not want to inbreed if it will harm next year's chicks. My hens are over six months old. I also have some thoroughbred Toulouse Geese which are related to one another. Should I change ganders this season? I used the same one last year but my geese are not as large this year as the ones I raised two years ago.—Mrs. R. L.

Answer—If your Plymouth Rock chickens are not full brother and sister you can mate them together if they are vigorous without inbreeding too closely. About the geese: I think that the reason your geese are not as large this year is that this year the mother geese were too young, or it may be that you did not feed them rightly.

THE Advertisers in this book are recommended by Mrs. Basley as being responsible and strictly reliable in their dealings. ¶ You will find them worthy of your patronage.

WHERE TO BUY BREEDING STOCK AND EGGS

WHERE TO BUY BREEDING STOCK AND EGGS